Oral Poetics and Cognitive Science

linguae & litterae

Publications of the School of Language & Literature
Freiburg Institute for Advanced Studies

Edited by
Peter Auer, Gesa von Essen, Werner Frick

Volume 56

Oral Poetics and Cognitive Science

Edited by
Mihailo Antović and Cristóbal Pagán Cánovas

DE GRUYTER

ISBN 978-3-11-034838-5
e-ISBN (PDF) 978-3-11-034853-8
e-ISBN (EPUB) 978-3-11-038468-0
ISSN 1869-7054

Library of Congress Cataloging-in-Publication Data
A CIP catalog record for this book has been applied for at the Library of Congress.

Bibliographic information published by the Deutsche Nationalbibliothek
The Deutsche Nationalbibliothek lists this publication in the Deutsche Nationalbibliografie; detailed bibliographic data are available on the Internet at: http://dnb.dnb.de.

Typesetting: jürgen ullrich typosatz, Nördlingen
Printing and binding: Hubert & Co. GmbH & Co. KG, Göttingen
♾ Printed on acid-free paper
Printed in Germany

www.degruyter.com

Contents

Preface – Acknowledgements

The editors thank the Freiburg Institute for Advanced Studies for awarding them the *Tandem Fellowship in Linguistics and Literary Studies*. Thanks to this splendid support, during the fall and winter 2012–13 we had the privilege of working at FRIAS on our joint project, entitled “Towards a Cognitive Oral Poetics: Traditional Epic and Cognitive Linguistics”. This was, as far as we know, the first project seeking to connect major approaches from cognitive linguistics, such as construction grammar or frame semantics, to the central tenets of oral poetics, mainly the research tradition on oral formulaic style originated by the Parry-Lord theory of composition in performance. Besides the connections that we detected between these fields, we were also inspired by other authors who have been building bridges between oral poetics and cognitive science, as well as by researchers in both disciplines who have worked independently, not realizing how relevant their work may be for the other field.

Fortunately for us, the many privileges associated with the FRIAS tandem fellowship also included support for organizing a conference on “Oral Poetics and Cognitive Science”[1], which took place at the Freiburg Institute for Advanced Studies from 24 to 26 January, 2013. This conference gave us the opportunity to bring together some of the researchers that had inspired our project, as well as other scholars whose work is very relevant for connecting oral poetic traditions with research on mind and language in general. To pave the way towards a cognitive oral poetics, a multidisciplinary group of experts examined the possibilities that the analytic tools from cognitive linguistics and cognitive science offer for the study of oral poetry. At the same time, we wanted to explore how oral poetics, especially in the empirical tradition originated by Parry and Lord, provides precious data for central issues in cognitive science and linguistics: language acquisition, the emergence of non-compositional meaning, idiomaticity, the origins of language and music, social cognition, the negotiation of meaning during performance, and many others.

The conference was also an invitation to imagine a cognitive oral poetics, a new field for the bi-directional exchange between two disciplines that have experienced a quick expansion during the past half-century or so: oral poetics and cognitive science. We had the privilege of enjoying interrelated talks by oral poeticians, cognitive scientists, linguists, and literary scholars, all dealing with related topics, and seeking to connect what their respective disciplines have to offer. Thanks to the generous hospitality of FRIAS, we also enjoyed long debates,

1 See https://sites.google.com/site/oralpoetcogsci/.

roundtables, and Q&A sessions. This book presents a selection of papers written after this experience. We thank all the participants at the conference, and our contributors in particular, for providing enthusiastic support and very useful feedback. We are especially grateful to Anna Bonifazi, alongside her PhD students Mark de Kreij and Annemieke Drummen, for their invaluable and insightful support over a long period, as well as for inviting us as visitors to their project at Heidelberg University. We thank Professors Egbert Bakker, Winfried Menninghaus, and Mark Turner for making room for our conference in their busy agendas.

For all their support during our research stay and conference in Freiburg, as well as throughout the edition of the volume, we feel extremely fortunate that our paths crossed with Coordinator of FRIAS Language and Literature School Gesa von Essen, FRIAS Linguistics Director Peter Auer, and Werner Frick, FRIAS Literature Director and Speaker of the Board of Directors at FRIAS. We would also like to thank all the members of the FRIAS community, who provided a thriving environment for our research, and especially the participants in the Linguistics and Literature Seminars. Finally, we thank the two anonymous reviewers of de Gruyter's *Linguae et Litterae* series for their work, which has allowed us to improve the volume in many ways.

Cristóbal Pagán Cánovas, University of Navarra and
Mihailo Antović, University of Niš

Introduction: Oral poetics and cognitive science

1 A new field about the oldest verbal art

This book is an invitation to reflect on the value of oral poetic traditions for the study of aesthetics, language, and the mind. It is also an invitation to explore how much more we can learn about orality in verbal art if we combine the analytic tools that cognitive science and oral poetics have been developing separately. Cognitive approaches to literature have become a burgeoning field over the last decades. Understanding the cognitive mechanisms that underlie verbal art requires what we could call a cognitive oral poetics. That is to say, an account of how the human mind produces the poetic in its original form and context: the oral performance situated within the tradition.

The variety of cultural traditions around the world magnificently reflects human diversity. But when it comes to preserving and studying verbal art, music, or plastic arts, we easily forget two obvious facts. One is that, no matter how fascinating their particularities may be, all traditions are products of the human mind. There lies their true intrinsic value, what makes each and every tradition a treasure, an invaluable part of the heritage of mankind. Languages, cultures, and nations come and go. As fascinating as their individual study can be, what matters most is what they can tell us about human nature.

The second fact is that oral traditions came first, a long, very long time before writing was invented and disseminated. Oral performances have been the only vehicles of art and knowledge—including plastic arts, inseparably linked to myth and performance in their origins—for over 90% of the fifty thousand or so years that cognitively-modern human beings have been around. Even today, barely half of the world's population masters reading and writing beyond the basics. To this we must add that oral traditions are intrinsically multiform and abundant (no publishing limitations), which makes them far more productive than written traditions. As a result, the vast majority of the poetry, music, and plastic arts being produced around the globe right now are situated within an oral milieu. Extremely rich and complex oral traditions still develop and flourish, sometimes interacting with literacy, sometimes not.

The combination of these two facts leads to a necessary conclusion: oral traditions are crucial for understanding the aesthetic phenomenon, culture, and the human mind. Indeed, literacy (not merely writing) was bound to change how we communicate, learn, even think... in dramatic and irreversible ways. But this was to happen only much later, after many millennia in which human beings were communicating, learning, and thinking without any extended use of writing. As invaluable as written or other forms of mediated communication are for us, they are not “the original”; they were not around in the beginnings of human cognition. Just like all other cultural manifestations, oral traditions from around the world have built on the same set of cognitive universals. But oral traditions bring us closer to the “natural exercise” of those basic mental abilities, that is, to the situation in which the human mind constructs a culture without texts, relying almost exclusively on the spoken and sung word. As a result, the many phenomena recurring across oral traditions, such as epic formulaic style, story-patterns, or infant-directed song, are “natural” products of our capacities for meaning construction and aesthetic experience. All these phenomena arise, in one way or another, from the necessities of live performance.

Notwithstanding all this, oral traditions currently receive very little attention, if any, from institutions and researchers devoted to the study of literature, music, arts, or language. Since research and academia are the products of literacy, their focus is, almost exclusively, on literate materials. The rapidly-growing interdisciplinary work connecting the humanities with the cognitive sciences has not changed this tendency, despite the higher “ecological validity” of oral traditions for studying human cognition. We would like to suggest that this situation should be rethought. As a contribution to this revision, and with no claims of being exhaustive or systematic, this volume points at some of the possible connections between cognitive science and oral poetics, showcasing the interdisciplinary research that can be done at the intersection of both disciplines. At the same time, the present book also shows that, after all, there are some exceptions to the general lack of attention paid to oral poetics from both literary studies and linguistics/cognitive science.

2 Oral poetics and the study of language and cognition

After centuries of the focus on written literature alone in the humanities, the Parry-Lord research on *oral composition in performance*—the online, improvisation-based, poem-producing technique of illiterate oral singers—conducted in the 1930s and 1950s on the territory of what was then Yugoslavia, represented arguably the major breakthrough in classics and oral tradition studies[1]. In the picture offered by Parry and Lord, oral poetry is anything but a 'recollection in tranquility' of the Wordsworthian kind. Quite the contrary, it exposes all major linguistic phenomena in a rather "stressful" situation: a live performance, where the singer is constantly under strong cognitive pressure to meet the 'demands' of his audience, so that they should not leave the coffee shop in boredom, ridding him of his fee or, more importantly, of future opportunities to perform. Like a jazz improviser thinking about his forthcoming harmonies as he plays the current themes, the oral singer is constantly in the position to "sing and think forward", unfolding the storyline events to come, in his virtual "simulation", likely multimodal in nature. This of course calls for a maximum cognitive efficiency, which in turn may reveal some "natural" phenomena related to language use much better than studying a written text, a well-rehearsed poetic performance, or even a plain, relaxed everyday conversation.

Parallel to that, the so called "cognitive revolution", which originated with the work of Noam Chomsky and followers in the 1950s, has probably been the most important movement cutting across all the sciences of the mind for the past one hundred years. In the last two decades we have witnessed a new wave of cognitive revolution, which some authors label "second generation cognitive science"[2] or even "postcognitivism"[3]. This new paradigm rejects some notions of the 1970s approaches to cognition, such as strict nativism, formal modeling based on computer metaphors, modularity of the mind or localizationist neuroscience. Rather, it insists on holism, connectionism, constructivism, an important role of our bodies in constructing mental concepts (*embodiment*), and the emergence of mental phenomena through the process of use. This paradigm shift in cognitive science has a number of important points of coincidence with the Parry-Lord approach: the idea that language learning and verbal creativity result from usage and performance, and build on general cognitive capacities and cultural context. On such a view, the basic units of a cognitive system, of language and oral poetry alike, are not sets of transformational or formal rules, but functional form-meaning pairs acquired through instance-based generalization.

---

1 Milman Parry, *Making of Homeric Verse: The Collected Papers of Milman Parry*. Ed. by Adam Parry. Oxford 1971; Albert Bates Lord, *The Singer of Tales*. Cambridge, MA 1960). See Pagán Cánovas and Antović (this volume) for an introduction to the Parry-Lord theory and for its connections with cognitive linguistics.

2 L. I. Qi-Wei, "Cognitive Revolution and Second-Generation Cognitive Science." *Acta Psychologica Sinica* 12 (2008): 1306–1328.

3 Brendan Wallace, A. J. Ross, J. B. Davies, and Tony Anderson, *The mind, the body and the world: Psychology after cognitivism?*. Exeter 2007.

Today, half a century after Parry and Lord's research, it seems to us that their approach to oral composition in performance may be revived, and their work appreciated even better as being ahead of its time, if it were viewed through the lenses of the cognitive sciences. For one thing, a common complaint by cognitive linguists against appreciating literary corpora is that literature is the product of too artificial a process to be of much interest in the study of natural language. The Parry-Lord theory of oral composition in performance refutes this argument rather easily: in oral poetic traditions there is no fixed text and no actual memorization. Traditional epic singers, whom today one can hardly find in Europe, but who were still very much active in Yugoslavia around World War Two, did not learn their songs by heart, but rather improvised on the basis of patterns which they had previously interiorized through imitation.

Therefore, the tradition was oral, creative, and very close to being "natural" in the sense required by modern linguists. One could imagine a young future singer following his father or role-model in coffee houses, listening to him perform the same or similar songs to the audience time and again, always with slight variations, also perhaps starting to practice a little bit at a very young age, and then becoming more and more proficient as time went by. This skill acquisition process reveals many phenomena that are potentially of interest for cognitive linguists. To start with, cognitive approaches to grammar originated from the study of idiomatic structures, long neglected in more formal approaches to language. The process of composition in performance is based precisely on enhanced idiomaticity. Singers constantly construct conventionalized form-meaning templates associated with narrative, structural, metrical, melodic, and prosodic conditions in which they are producing their songs: hence the connection with "traditional" domains of interest in the language science, from phonology to semantics. Further, cognitive linguistics is interested in "natural" language acquisition, that which is not based on much explicit instruction. Composition in performance likewise involves no writing or systematic coaching, which is quite analogous to first language acquisition, or even "traditional" second language acquisition, in which a young learner is simply "immersed" in a new environment and left to acquire the new tongue.

For a further connection, cognitive linguists insist that language is not acquired through a "universal grammar", an influential, but never quite corroborated theoretical construct of generative linguistics, standing for human specialized genetic capacity to effortlessly master a first language. Rather, cognitivists believe that language emerges from use, in a much more piecemeal, constructivist fashion, driven by instinctive pattern finding, statistical elimination and lots of unconscious, Gestalt-based generalization. Composition in performance could become an additional pillar to strongly support constructivism in language acquisition, in that it is also a process in which an elaborate verbal skill is acquired through instance-based generalizations emerging from usage, where transformational rules are secondary, if any. A generative grammar would thus hardly be able to explain composition in performance (and probably not much care about it, as it mostly focuses on *competence* as opposed to *performance*, albeit the latter term here is used in a much more technical sense than Parry and Lord would have had in mind). The argument is important, since oral poetry, including the formulaic style of traditional epic, is so widespread across cultures and in historical time—indeed so prevalent among human communities—that modern cognitive sciences would overlook a significant segment of human creativity if the phenomenon of composition in performance were neglected. In turn, if this product of oral creativity were better explicable by the epistemology of cognitive than generative linguistics, that might attract some additional attention for the former in the cognitive sciences in general.

Besides the connections with cognitive linguistics, oral poetics can be relevant to other linguistic approaches that highlight the conceptual and multimodal complexity involved in oral communication. As several contributions to this volume show, oral poetics can fruitfully combine with interactional linguistics, and especially with conversation analysis, as well as with recent developments in pragmatics and discourse analysis, which highlight phenomena that are often neglected when the focus is exclusively on written texts: turn-taking, repetition, prosody, discourse markers, etc.

Indeed, the connections between oral poetics and cognitive science need not be limited to the study of the verbal modality. An act of oral performance is an emergent, multimodal, intersubjective event of oral communication, always well integrated within a cultural tradition. It often involves singing, storytelling, playing a simple instrument such as the South Slavic *gusle*, gesticulating, making pauses while moving around in the performance setting, and requesting verbal and nonverbal responses from the audience. As such, the skill could be of interest in the most recent efforts in cognitive science and linguistics to encompass the full complexity of language and discourse by studying verbal and non-verbal communication from a multimodal perspective, integrating the study of language, co-speech gesture, bodily movement, visuo-spatial focus (as in camera panning or zooming in a TV broadcast), etc.[4]

---

4 See, e.g. Francis Steen and Mark B. Turner, "Multimodal Construction Grammar". In: Michael Borkent, Barbara Dancygier, and Jennifer Hinnell (eds.), *Language and the Creative Mind*. Stanford, CA 2013, http://papers.ssrn.com/abstract=2168035; David McNeill, *Gesture and Thought*. Chicago 2008); Gabriella Vigliocco, Pamela Perniss, and David Vinson, "Language as a Multimodal Phenomenon: Implications for Language Learning, Processing and Evolution". *Philosophi-*

In relation to this claim, as this text is being written, a massive transition in corpus-based cognitive linguistic studies seems to be underway, where the traditional written corpora are replaced by their multimodal counterparts, searchable databases of audiovisual materials, revealing not only text, but also speech, intonation, gesture, facial expression, and bodily movement. This is the case with the NewsScape Library of News Broadcasts, with over 250,000 hours of television made fully searchable as a written corpus, through close-captioning or subtitles synchronized with their corresponding images.[5] In a development of which they may not have even dreamed before, oral poetics scholars could now explore how resources such as the Milman Parry Collection of Oral Literature[6] or the World Oral Literature Project[7] could benefit from these technologies, giving rise to oral poetic archives as accessible and searchable for linguists as the most up-to-date multimodal corpora.

3 Cognitive poetics and oral poetry

Over the last three decades, cognitive science and the traditional humanities have developed many connections, mainly based on their shared interest in the mental processes underlying literature and the arts. In particular, cognitive scientific research on figurative language and linguistic creativity in general has facilitated the emergence of cognitive poetics as a vibrant field,[8] building on findings from

cal Transactions of the Royal Society B: Biological Sciences 369, no. 1651 (September 19, 2014): 20130292; S. Marlar Lwin, "Capturing the Dynamics of Narrative Development in an Oral Storytelling Performance: A Multimodal Perspective". *Language and Literature* 19.4 (2010): 357–377.

5 See (http://newsscape.library.ucla.edu/), hosted by the UCLA Library and developed by the Distributed Red Hen Lab for research on multimodal communication: https://sites.google.com/site/distributedlittleredhen/.

6 See http://chs119.chs.harvard.edu/mpc/.

7 Mark Turin, "Orality and Technology, or the Bit and the Byte: The Work of the World Oral Literature Project". *Oral Tradition* 173 (2013): 186; Mark Turin, Claire Wheeler, and Eleanor Wilkinson (eds.), *Oral Literature in the Digital Age: Archiving Orality and Connecting with Communities*. Cambridge 2013, http://www.openbookpublishers.com/product/186.

8 Some of the main overviews: Margaret H. Freeman, "Cognitive Linguistic Approaches to Literary Studies: State of the Art in Cognitive Poetics". In: *The Oxford Handbook of Cognitive Linguistics*, 2007, 1175–1202; Geert Brône and Jeroen Vandaele, *Cognitive Poetics: Goals, Gains and Gaps*. Berlin and New York 2009; Patrick Colm Hogan, *Cognitive Science, Literature, and the Arts: A Guide for Humanists*. London 2003); Elena Semino and Jonathan Culpeper, *Cognitive Stylistics: Language and Cognition in Text Analysis*. Amsterdam 2002.

cognitive linguistics,[9] as well as from other backgrounds, such as psycholinguistics.[10] However, a true interdisciplinarity between poetics and cognitive science still constitutes a big challenge.[11]

Cognitive poetics encounters at least two major difficulties that hinder its impact on cognitive science. One is that the (often necessary) interpretive speculation of literary studies is perceived as a lack of empirical rigor by cognitive scientists, who demand more tractable data. The other is, as we noted, that cognitive poeticians, whose background comes mainly from literary criticism, are almost exclusively interested in what many linguists regard as the most "artificial" manifestations of verbal creativity, namely, written texts resulting from careful elaboration, usually belonging to a very selective literary canon, and thus remote from everyday oral interactions, which take place in "natural" settings and in real-time communication.

A cognitive oral poetics could easily address both objections. Oral poetics works with records of authentic performances and, whenever possible, by directly witnessing the real performances themselves. Orally composed poetry is a "natural" practice recurring across cultures. It structures most social, everyday-life activities in predominantly oral societies, that is, the societies in which almost all humans lived until the second half of the twentieth century. Also, oral poetics, and particularly the research tradition in oral composition in performance originated by Parry and Lord, has always combined rigorous philological and historical scholarship with standard methods of data gathering in the social sciences and linguistics, such as anthropological fieldwork or corpus linguistics. On the other hand, scholars in oral poetics have shown that these empirical methods are indispensable for the study of the intrinsically aesthetic phenomenon called oral poetry. Thus a cognitive oral poetics can also address a major objection of literary scholars to cognitive linguists, namely, that their linguistic analyses do not account for what mainly concerns literary studies: artistic value, cultural tradition, particularities of style, or poetic effects. Incorporating the so far neglected oral studies offers the opportunity to refound cognitive poetics as a discipline

---

9 Mark Turner, *Death Is the Mother of Beauty: Mind, Metaphor, Criticism.* Chicago 1987; Mark Turner, *Reading Minds: The Study of English in the Age of Cognitive Science.* Princeton 1991; Mark Turner, *The Literary Mind.* New York 1996; George Lakoff and Mark Turner, *More than Cool Reason: A Field Guide to Poetic Metaphor. Chicago* 1989; Raymond W. Gibbs, *The Poetics of Mind: Figurative Thought, Language, and Understanding.* Cambridge 1994.

10 Reuven Tsur, *Toward a Theory of Cognitive Poetics.* Brighton and Portland 2008; David S. Miall, *Literary Reading: Empirical & Theoretical Studies.* New York et al. 2006.

11 M. J. Bruhn, "Introduction: Exchange Values: Poetics and Cognitive Science". *Poetics Today* 32.3 (2012): 403–60.

including two equally important fields of study, the oral and the written, where both forms of expression should be given equal attention, if necessary also in a contrastive or comparative fashion.[12]

The regularities observed across distant oral traditions expose poetic universals,[13] but they can be even more valuable if considered within the broader frameworks of linguistics and cognitive science. In order to achieve its full impact, comparative oral poetics needs to be grounded in the study of the mind, just like any field investigating regularities across literary traditions.[14]

The mental feats accomplished by epic singers constitute a fascinating phenomenon for the study not only of memory,[15] but also of idiomaticity, conceptual structure, and many other key issues in language and cognition, especially if we focus on the common operations of meaning construction shared by oral traditions around the world. By manipulating these operations to serve the strong communicative demands of performance, oral poetry exposes cognitive patterns more clearly than everyday language does. However, so far there has been no systematic account of oral poetics within a cognitive scientific framework.

4 The contributions to this volume

The exploratory nature of our project has naturally produced an exploratory volume, in which a variety of researchers point at different paths that cognitive oral poetics could follow. Such a volume does not easily accept division into sections. But we did organize the chapters on the basis of their main thematic connections. The first three chapters (Minchin, Person, Duffy & Short) study oral poetry in comparison with everyday language, including conventional metaphors and folk models. The second two (Pagán Cánovas & Antović, Boas) focus on oral formulaic style, connecting it to cognitive linguistic research: construction grammar and frame semantics. The last three papers (Bonifazi, de Kreij, Zeman) are related to one of the most promising fields in oral poetics right now: the multimodal online

---

12 See Walter J. Ong, *Orality and Literacy*. London and New York 1982.

13 Claudio Guillén, *The Challenge of Comparative Literature*, 69–70, 173–174. Harvard 1993.

14 As argued for literary studies in general by P. C. Hogan, "Literary Universals". *Poetics Today* 18 (1997): 223–249; Hogan, *Cognitive Science, Literature, and the Arts*; Turner, *Reading Minds*; Turner, *The Literary Mind*.

15 David C. Rubin, *Memory in Oral Traditions: The Cognitive Psychology of Epic, Ballads, and Counting-out Rhymes*. (Oxford University Press US, 1995); Elizabeth Minchin, *Homer and the Resources of Memory: Some Applications of Cognitive Theory to the Iliad and the Odyssey*. Oxford 2001).

structuring of oral poetic discourse, which guides the audience's mental simulation of the story through particles, formulaic patterns, and other formal resources.

Elizabeth Minchin studies formal, conceptual, and narrative repetition in Homeric epic, and concludes that it serves the main strategies of repetition in spontaneous conversation (production, comprehension, connectedness, and interaction), albeit in ways that show the creative mind of the poet at work under the great cognitive stress of performance. *Raymond Person* compares ring composition, one of the most typical devices of oral poetry, across Serbo-Croatian epic, Homer, *Beowulf*, and *Genesis* with selected examples from conversation, showing that the epic usage is an expansion of the cognitive operations that we find in everyday interactions. *William Duffy & William Short* study the conventional metaphors underlying ancient Greek language about the epic tradition, and propose that, through conceptual metaphors, Greek speakers separated composition from performance and created an artificial—though useful—perception of the oral tradition as something fixed and stable.

Cristóbal Pagán Cánovas & Mihailo Antović compare formulas and constructions, the central theoretical constructs of the Parry-Lord theory of composition in performance and cognitive grammar, showing how both concepts are based on the same view of linguistic knowledge as a result of instance-based generalization, and indicating some of the possibilities for combining the two approaches. From a cognitive linguistic perspective, *Hans Boas* describes frame semantics and construction grammar and proposes how these theories, also including the computational resources of FrameNet, can be used to analyze poetic sources.

The emphasis on the pragmatic and functional values of formulas has produced very relevant work on deixis[16] and discourse markers,[17] as devices for narrative structuring and story segmentation, as well as resources for guiding the listener's mental simulation, for zooming in on / out of characters and events, and other effects related to the mental imagery underlying the multimodal aspects of meaning construction.[18] The Greek particles project led by Anna Bonifazi at the

16 See especially Egbert J. Bakker, *Pointing at the Past: From Formula to Performance in Homeric Poetics*. Washington DC 2006.

17 See especially Anna Bonifazi, *Homer's Versicolored Fabric: The Evocative Power of Ancient Greek Epic Wordmaking*. Washington DC 2012.

18 David F. Elmer, "Presentation Formulas in South Slavic Epic Song". *Oral Tradition* 24.1 (2010); Anna Bonifazi and David Elmer, "Composing Lines, Performing Acts: Clauses, Discourse Acts, and Melodic Units in a South Slavic Epic Song". In: Elizabeth Minchin (ed.), *Orality, Literacy and Performance in the Ancient World*. Leiden 2011; Anna Bonifazi, "Memory and Visualization in Homeric Discourse Markers". In: Anne Mackay (ed.), *Orality, Literacy, Memory in the Ancient Greek and Roman World*, 35–64. Leiden and Boston 2007.

University of Heidelberg[19] has led to pioneering research along these lines, well represented by the contributions to this volume by Mark de Kreij and Bonifazi herself. This is a promising research line that has great potential connections with very recent work in multimodal construction grammar.[20] The "multimodality section" of this volume reflects this burgeoning field.

Anna Bonifazi proposes a pragmatic reading of vocal discontinuities (in a Serbo-Croatian epic song) and various discourse markers and particles (in both Greek and Serbo-Croatian), analyzing the multimodal, metanarrative functions of particles in both traditions, and identifying their role as cues to structuring. *Mark De Kreij* studies left-dislocation in Homeric Greek from a discourse perspective, showing that it (re)establishes contextual frames or accommodates a change of viewpoint, therefore reflecting the journey of performer and audience through the storyworld. Finally, *Sonja Zeman* analyzes perspectivization and Theory of Mind in metanarrative devices in Middle High German (semi-)oral poems, providing a link to the relation between orality and literacy and reflecting on the comparative methodologies needed by a cognitive oral poetics.

References

Bakker, Egbert J. *Pointing at the Past: From Formula to Performance in Homeric Poetics*. Washington, DC 2006.

Bonifazi, Anna. "Memory and Visualization in Homeric Discourse Markers." In: Anne Mackay *(ed.), Orality, Literacy, Memory in the Ancient Greek and Roman World*, 35–64. Leiden and Boston 2007.

Bonifazi, Anna. *Homer's Versicolored Fabric: The Evocative Power of Ancient Greek Epic Wordmaking*. Washington, DC 2012.

Bonifazi, Anna and David Elmer. "Composing Lines, Performing Acts: Clauses, Discourse Acts, and Melodic Units in a South Slavic Epic Song." In: Elizabeth Minchin *(ed.), Orality, Literacy and Performance in the Ancient World*. Leiden 2011.

Brône, Geert and Jeroen Vandaele. *Cognitive Poetics: Goals, Gains and Gaps*. Berlin and New York 2009.

Bruhn, Mark J. "Introduction: Exchange Values: Poetics and Cognitive Science". *Poetics Today* 32.3 (2012): 403–460.

Elmer, D. F. "Presentation Formulas in South Slavic Epic Song". *Oral Tradition* 24.1 (2010): 41–59.

---

19 See http://www.uni-heidelberg.de/fakultaeten/philosophie/skph/emmy_noether/index.html.

20 Francis Steen and Mark B. Turner, "Multimodal Construction Grammar". In: Michael Borkent, Barbara Dancygier, and Jennifer Hinnell (eds.), *Language and the Creative Mind*. Stanford, CA 2013; Elisabeth Zima, "Gibt es multimodale Konstruktionen? Eine Studie zu [V (motion) in Circles] und [all the Way from X PREP Y]". *Gesprächsforschung – Online-Zeitschrift zur verbalen Interaktion* 15 (n.d.) (2014): 1–48.

Freeman, Margaret H. "Cognitive Linguistic Approaches to Literary Studies: State of the Art in Cognitive Poetics." In: Dirk Geeraerts and Hubert Cuyckens (eds.), *The Oxford Handbook of Cognitive Linguistics*, 1175–1202. Oxford 2007.
Gibbs, Raymond W. *The Poetics of Mind: Figurative Thought, Language, and Understanding*. Cambridge 1994.
Guillén, Claudio. *The Challenge of Comparative Literature*. Harvard 1993.
Hogan, Patrick C. *Cognitive Science, Literature, and the Arts: A Guide for Humanists*. New York 2003.
Hogan, Patrick C. "Literary Universals". *Poetics Today* 18.2 (1997): 223–49.
Lakoff, George and Mark Turner. *More than Cool Reason: A Field Guide to Poetic Metaphor*. Chicago 1989.
Lord, Albert Bates. *The Singer of Tales*. Cambridge, MA 1960.
Marlar Lwin, S. "Capturing the Dynamics of Narrative Development in an Oral Storytelling Performance: A Multimodal Perspective". *Language and Literature* 19.4 (2010): 357–377.
McNeill, David. *Gesture and Thought*. Chicago 2008.
Miall, David S. *Literary Reading: Empirical & Theoretical Studies*. New York 2006.
Minchin, Elizabeth. *Homer and the Resources of Memory: Some Applications of Cognitive Theory to the Iliad and the Odyssey*. New York 2001.
Ong, Walter J. *Orality and Literacy*. London and New York 1982.
Parry, Milman. *The Making of Homeric Verse: The Collected Papers of Milman Parry*. Ed. by Adam Parry. Oxford 1971.
Rubin, David C. *Memory in Oral Traditions: The Cognitive Psychology of Epic, Ballads, and Counting-out Rhymes*. New York 1995.
Semino, Elena and Jonathan Culpeper. *Cognitive Stylistics: Language and Cognition in Text Analysis*. Amsterdam 2002.
Steen, Francis and Mark Turner. "Multimodal Construction Grammar". In Michael Borkent, Barbara Dancygier, and Jennifer Hinnell *(eds.), Language and the Creative Mind*. Stanford, CA 2013.
Tsur, Reuven. *Toward a Theory of Cognitive Poetics: Second, Expanded and Updated Edition*. Brighton and Portland 2008.
Turin, Mark. "Orality and Technology, or the Bit and the Byte: The Work of the World Oral Literature Project". *Oral Tradition* 28.2 (2013): 173–186.
Turin, Mark, Claire Wheeler, and Eleanor Wilkinson (eds.). *Oral Literature in the Digital Age: Archiving Orality and Connecting with Communities*. Cambridge 2013.
Turner, Mark. *The Literary Mind*. New York 1996.
Turner, Mark. *Reading Minds: The Study of English in the Age of Cognitive Science*. Princeton 1991.
Turner, Mark. *Death Is the Mother of Beauty: Mind, Metaphor, Criticism*. Chicago 1987.
Vigliocco, Gabriella, Pamela Perniss, and David Vinson. "Language as a Multimodal Phenomenon: Implications for Language Learning, Processing and Evolution". *Philosophical Transactions of the Royal Society B: Biological Sciences* 369, no. 1651 (September 19, 2014): 20130292.
Zima, Elisabeth. "Gibt es multimodale Konstruktionen? Eine Studie zu [V (motion) in Circles] und [all the Way from X PREP Y]." *Gesprächsforschung – Online-Zeitschrift zur verbalen Interaktion* 15 (2014): 1–48.

Elizabeth Minchin, The Australian National University

Repetition in Homeric epic: Cognitive and linguistic perspectives

Abstract: Why are there so many instances—and so many varieties—of repetition in the oral epics that we associate with Homer's name? In this paper I begin with Deborah Tannen's observations on repetition in spontaneous conversational discourse. I then turn to the oral epic tradition that had flourished in early Greece to consider how repetition operates in its two representative poems, the *Iliad* and the *Odyssey*. In studying the epic poet's repetition of type-scenes, his use of repeated elements in the telling of stories, and his use of both parallelism and repetition in similes, I draw some conclusions about how poets in this oral tradition manipulated this device in order to meet the heavy demands of performance before a "live" audience—and how those same repeated elements supported their listeners as they attempted to process the tale.

I find that repetition in oral traditional storytelling serves the functions of production, comprehension, connectedness, and interaction—the very functions that Tannen has identified in spontaneous conversation. But, as we shall see, the ways in which—and the degree to which—these functions are served are in some respects different. The highly formalized patterns of usage that we observe in these records of oral traditional epic offer all scholars who are interested in cognition an unusually clear view of a creative mind at work.

1 Introduction

Repetition in everyday discourse is a phenomenon with which we are all familiar: in conversation we repeat our own words and the words of others; we readily identify repetition when we hear it. Such repetition may derive from expediency; it may derive from consideration for one's audience, or a desire to get one's message across; it may be used for humorous effect; it may indicate insolence. In a useful paper Deborah Tannen identifies four functions of repetition in spontaneous speech, illustrating these functions with examples recorded in informal contexts.[1] The first of these functions is *production*. When a new speaker chooses

1 Deborah Tannen, "Repetition in Conversation: Towards a Poetics of Talk". *Language* 63 (1987): 574–605.

to repeat some of his or her own words, or those of a previous speaker, he or she then produces speech more fluently, using ready-made language in a more efficient, faster, and "less energy-draining" way.[2] The second function is *comprehension*. When speakers repeat words and phrases that are already familiar, they create discourse that is less dense. The listener, therefore, is under less pressure as he or she absorbs the new information. These first functions address cognitive issues; we might say that speakers find repetition useful, both on their own account and with their listeners in mind, in recognition of the relatively low "cognitive ceiling" of both parties, since speakers and listeners are operating in real time.[3] The third function of repetition is *connection*. Speakers use repetition to serve a referential or tying function, which may also be evaluative. The fourth function is *interaction*, whereby the participants' social goals are fulfilled—when one speaker's words, for example, are picked up by another, the participants are linked not only to the discourse but to each other. Tannen, indeed, puts greater weight on these last two functions. From her perspective, repetition serves a need for interpersonal involvement, an emotional experience of rapport, or connectedness. And this, she argues, underpins one of the key criteria of language in use—*coherence*.

Building on the observation of Paul Friedrich that all language is poetic in varying degrees and that conversation may be considered as "rough drafts for poetry",[4] Tannen argues that we find repetition in two social contexts: in conversation, where it is "spontaneous, pervasive and often relatively automatic", and in the written texts of literary discourse, where everyday practices are "artfully developed and intensified".[5] In this latter context Tannen (unsurprisingly) bypasses the functions of repetition that facilitate *production* and *comprehension*, and considers only its interactive ends. In poetic texts and declamatory contexts, she notes, repetition is considered a useful and elegant means by which poets or speakers can achieve their goals, or, in rhetorical terms, can make their point.[6] In this regard, Tannen demonstrates, persuasively, that ordinary conversation and literary discourse are more closely linked than might previously have been thought.

---

2 Ibid., 581.

3 Douglas Biber, *Variation across Speech and Writing*. Cambridge 1988, 163.

4 Paul Friedrich, *The Language Parallax: Linguistic Relativism and Poetic Indeterminacy*. Austin 1986, 3, 33.

5 Tannen, "Repetition in Conversation", 580–581.

6 Cf. Jeanne Fahnestock, *Rhetorical Style: The Uses of Language in Persuasion*. Oxford 2011, 230–231.

Tannen has made an important observation. But I note a significant omission from her study: there is no recognition of discourse at an intermediate stage along the oral-literate continuum. What has been overlooked is the kind of oral discourse that we find not in the informal everyday context but in more formal, even ritual, contexts.[7] One such context in the Ancient Greek world must have been the performance of heroic song, which, in time, culminated in the *Iliad* and the *Odyssey*. These poems, by tradition associated with Homer's name, derive from an oral tradition that had developed many centuries before the introduction of alphabetic writing. Thanks to the important ethnographic work of Milman Parry and Albert Lord, who studied oral poets in the then living Serbian and Croatian traditions, we can be confident that the oral poet in the ancient Greek world performed without reference to, or the support of, a written text; and that he had not memorized his song in advance of his performance: rather, he composed, or improvised, *as he performed*.[8]

As Wallace Chafe points out in his discussion of discourse associated with formal occasions (such as traditional epic song), the very fact that material of this kind is held in high regard assures its survival over an extended period of time.[9] These two factors in combination (esteem and longevity) influence the linguistic choices a poet will make. That is, he will respond to the formality of the occasion and the sense of responsibility he feels by using a more formal register. And, of course, we must acknowledge that there will have been, from performance to performance, repeated opportunities for the poet to rehearse and review.[10] It should not surprise us, therefore, that because of these factors and these opportunities, the language of ritual (and of oral literature) is more formalized and more polished than the language of every day: Egbert Bakker, for example, describes the language of repeated formulas that characterizes Homeric epic as 'special'.[11]

Oral traditional song, in fact, has much in common with everyday conversation: it is oral; it is directly interactive; and it does not use a written text as an aid. In these very respects, however, oral song imposes far greater demands on both the poet and his audience than everyday talk imposes on a group of speakers. Let me begin with the poet: first, fluency is a non-negotiable requirement (breakdown

---

7 Wallace Chafe, "Integration and Involvement in Speaking, Writing, and Oral Literature". In Deborah Tannen (ed.), *Spoken and Written Language: Exploring Orality and Literacy*, 35–53, 49–52. Norwood, N.J. 1982.

8 Adam Parry (ed.), *The Making of Homeric Verse: The Collected Papers of Milman Parry*. New York and Oxford 1971/1987; Albert B. Lord, *The Singer of Tales*. Cambridge, MA 1960.

9 Chafe, "Integration and Involvement".

10 Ibid., 49–52.

11 Egbert J. Bakker, *Poetry in Speech: Orality and Homeric Discourse*. Ithaca and London 1997.

would bring a damaging loss of face); second, oral performance is monologic (there is no sharing of the floor with another speaker); and, third, in the case of epic song (such as substantial portions of the great epics under discussion), the performance may continue for quite some time. On the other hand, as far as the audience is concerned, the limitations of memory and attention that constrain a poet's ability to perform constrain them also as they process the song.

Oral traditional poetry, as we shall observe, also has elements in common with literary discourse; but in some respects it is significantly different. The literary texts with which we are familiar today have been composed with the aid of writing (although I do not doubt that a measure of oral input is also detectable); they have been composed (in most cases) over a far longer period of time (the poet has been able to ponder word choice and language strategy at leisure); and an audience of readers (unlike listeners) may choose how and when they are to engage with the text. They may, for example, skim quickly over the text at one sitting, or they may read it slowly, absorbing all its nuances. Repetition in literary texts rarely serves the functions of production and comprehension discussed above: and yet these functions, as I shall show, are critical to the oral poet and to members of his audience.

I propose therefore to take Tannen's discussion one step further. I shall examine repetition—as well as its close relative parallelism—as it occurs in these ancient oral performance 'texts', to observe how that ancient oral tradition employed this strategy: that is, I shall observe the functions of repetition in the more formalized discourse of oral poetry and relate them to both the functions that Tannen has identified in everyday discourse and the narrower range of functions that she finds in the discourse that we associate with written composition. Essential to my discussion will be the notion of cognitive capacity: how an oral poet and a typical member of his (listening) audience can operate successfully within the limitations of memory.[12] As we observe the ways in which a poet in this ancient tradition used repetition to manage the business of storytelling, we cannot help but learn more about cognition itself.

---

12 Elizabeth Minchin, "Poet, Audience, Time, and Text: Reflections on Medium and Mode in Homer and Virgil". In Ruth Scodel (ed.), *Between Orality and Literacy: Communication and Adaptation in Antiquity. Orality and Literacy in the Ancient World X*, 267–288. Leiden 2014.

2 Repetition in Homeric epic

I begin with a brief but necessary account of the language of the epics. There has been much discussion of repetition at the level of the small phrasal unit that is identified as the Homeric formula.[13] These units are in many cases heavily used in the epics—and are far more intensively employed in these texts than are comparable phrases in everyday talk or, indeed, in literary composition. Indeed, Mueller observes that whereas literary works of all eras, from Plato's *Republic* to Jane Austen's *Emma* to Marx's *Das Kapital*, do not differ in their threshold value for repetition (the point at which a repeated word or phrase is no longer overlooked, or taken for granted), Homeric poetry consistently violates this threshold value.[14] This is understandable, since the formula (often a noun qualified by an epithet) was developed precisely in order to facilitate the composition of metrical storytelling in an oral context; it was designed to be used and used again. So the oral poet does not avoid (as we might) repeating the phrase Φοῖβος Ἀπόλλων (*Phoibos* Apollo) at *Iliad* 1. 43, 64, 72, 182; πόδας ὠκὺς Ἀχιλλεύς (*swift-footed* Achilles) at *Iliad* 1. 58, 84, 148, 215, 364; or κρείων Ἀγαμέμνων (*lordly* Agamemnon) at *Iliad* 1. 102, 130, 285.

But what impact might such prefabricated units of speech have on an audience-member? The performance of oral traditional epic is clearly different from the experience of storytelling in everyday talk. The poet's language, marked by noun-epithet formulas, is certainly distinctive. Every hero is described as remarkable, every artefact as admirable or exquisite. The poet thereby creates a sense of occasion, as the listener is transported into another world.[15] What is more important for my argument here is that formulaic language of this kind eases the task of listening. Not only does the formula take somewhat longer to enunciate than does a noun unadorned, it also has the advantage of being repeated material. The formulaic phrase slows the presentation of new information and allows listeners an extra moment to absorb it: indeed, capacity in short-term memory is freed up so that the listener can focus on the storyline and absorb the poet's evaluative cues.[16]

---

13 Joseph Russo, "Formula". In: Margalit Finkelberg (ed.), *The Homer Encyclopedia*, vol. 1, 296–298. Chichester 2011.

14 Martin Mueller, "Repetitions". In: Margalit Finkelberg (ed.), *The Homer Encyclopedia*, vol. 3, 739–742. Chichester 2011.

15 Joseph Russo, "Sicilian Folktales, Cognitive Psychology, and Oral Theory". In: Thomas M. Falkner, Nancy Felson, David Konstan (eds.), *Contextualizing Classics: Ideology, Performance, Dialogue*, 151–171, 168–169. Lanham 1999.

16 Deborah Tannen, "Oral and Literate Strategies in Spoken and Written Narratives". *Language* 58 (1982): 1–21; Elizabeth Minchin, *Homer and the Resources of Memory: Some Applications of Cognitive Theory to the* Iliad *and the* Odyssey, 123–124. Oxford 2001.

I should make it clear at this point that this "moment" may be measured only in fractions of a second; but even so brief a period is as welcome to the listener as it is to the singer.

My principal concern in this paper, however, is with repetitions at a larger, structural level. I shall explore serial repetition in a suite of type-scenes, repetitions in stories told within the larger narrative, and repetition in similes—in which the parallelism between the simile and its corresponding narrative elements may also be counted as a form of repetition. I am interested not only in the various ways in which the tradition has standardized these discourse features for the convenience of poet and audience but also in its exploitation of repetition as part of this standardization.

Repetition in type-scenes. It was Milman Parry and Albert Lord who first saw the significance of type-scenes—repeated event sequences—as tools of performance in the context of oral epic storytelling. But, although it has been understood for some decades that an oral singer draws down such scenes from memory, what has been in doubt for some time has been precisely what sort of remembering is implied. Indeed, although Lord at several points speaks of the process of oral composition, he is remarkably silent on the role of memory: for example, the singer "comes to know" the details of the song; he is "equipped with" a store of formulas and themes and a technique of composition.[17]

I have proposed elsewhere that the evidence of the texts, along with findings in cognitive psychology, allows us to conclude that the type-scenes of oral traditional poetry (and, in our case, specifically of Homeric epic) are a cognitive phenomenon.[18] Each type-scene is an expression of a script that is stored in memory; each script stores knowledge in episodic form about everyday activities and routines—even, I argue, speech routines.[19] It is not surprising, therefore, that within each long epic we find that some scripts are referred to again and again, almost always in a carefully standardized form, as warriors arm for battle, as a hero undertakes a journey, or as a visitor is welcomed into the dwelling of a ruler, or, if we turn from scripts for action to scripts (or, as I call them, formats) for speech, as one hero rebukes another on the battlefield, or as a potential guest declines an invitation.[20] Poets in this tradition have reduced each event-sequence to certain essentials that are regularly instantiated. In an arming-scene, for example, the sequence in which a hero dons his armour will always be the same (greaves, corslet, sword, shield, helmet, and spear). This order is not haphazard; rather, it

17 Lord, *Singer of Tales*, 68, 99.

18 Minchin, *Homer and the Resources of Memory*, 32–72.

19 Elizabeth Minchin, *Homeric Voices: Discourse, Memory, Gender*, 10–12. Oxford 2007.

20 Ibid., 23–51, 52–73.

represents the sequence in which items of armour are necessarily donned. The routine nature of the script is not in itself remarkable; its completeness and its regularity is. I should add that the poet individualizes each arming-scene by offering a description of a selected item of armour, and this may or may not also include a story about that item's origins. The passage below describes Paris's actions as he arms for his contest with Menelaos, lingering briefly over his corslet:

κνημῖδας μὲν πρῶτα περὶ κνήμῃσιν ἔθηκε
καλάς, ἀργυρέοισιν ἐπισφυρίοις ἀραρυίας·
δεύτερον αὖ θώρηκα περὶ στήθεσσιν ἔδυνεν
οἷο κασιγνήτοιο Λυκάονος· ἥρμοσε δ' αὐτῷ.
ἀμφὶ δ' ἄρ' ὤμοισιν βάλετο ξίφος ἀργυρόηλον
χάλκεον, αὐτὰρ ἔπειτα σάκος μέγα τε στιβαρόν τε·
κρατὶ δ' ἐπ' ἰφθίμῳ κυνέην εὔτυκτον ἔθηκεν
ἵππουριν· δεινὸν δὲ λόφος καθύπερθεν ἔνευεν·
εἵλετο δ' ἄλκιμον ἔγχος, ὅ οἱ παλάμηφιν ἀρήρει.

First he placed along his legs the fair greaves linked with
silver fastenings to hold the greaves at his ankles.
Afterwards he girt on about his chest the corslet
of Lykaon his brother since this fitted him also.
Across his shoulders he slung the sword with the nails of silver,
a bronze sword, and above it the great shield, huge and heavy.
Over his powerful head he set the well-fashioned helmet
with the horse-hair crest, and the plumes nodded terribly above it.
He took up a strong-shafted spear that fitted his hand's grip.

Homer, *Iliad* 3. 330–338 (transl. Lattimore)

In a further four passages in the *Iliad* the poet follows the same sequence: *Iliad* 11. 15–55; 15. 479–482; 16. 130–154; 19. 364–424.

On the other hand, if a hero departs on a journey, from his own residence or the palace of his host, the journey-scene is represented through the following elements: yoking of horses (the extra burden is loaded); mounting the chariot; whipping up the horses; their readiness to run; the route; and the destination. We should note that the poet has the option of elaborating on each element (compare the lengthy yoking element at *Iliad* 24. 266–278 with the slimmed-down account at 24. 278–279). Here is an economical version of this type-scene:

ἵππους τε ζεύγνυντ' ἀνά θ' ἅρματα ποικίλ' ἔβαινον·
ἐκ δ' ἔλασαν προθύροιο καὶ αἰθούσης ἐριδούπου·
μάστιξεν δ' ἐλάαν, τὼ δ' οὐκ ἀέκοντε πετέσθην.
ἷξον δ' ἐς πεδίον πυρηφόρον, ἔνθα δ' ἔπειτα
ἦνον ὁδόν· τοῖον γὰρ ὑπέκφερον ὠκέες ἵπποι.

> They yoked the horses again and mounted the chariots bright with
> bronze, and drove them out the front door and the echoing portico,
> and he whipped them into a run and they winged their way unreluctant.
> They came onto the plain full of wheat, and by that way
> made good their journey as their fast horses took this by-way.
>
> Homer, *Odyssey* 3. 492–496 (transl. Lattimore)

We see similar sequences at *Odyssey* 6. 72–84; 15. 144–146 with 182–184, 189–193, as well as the extended sequence referred to above, at *Iliad* 24. 266–350.

The oral poet in this tradition can use type-scene repetition in more sophisticated ways as well. At other points of the *Iliad* and the *Odyssey* we find repeated type-scenes presented as tightly-linked sequences in order to structure the narrative: see, for example, the sequence of contests that underpin the funeral games of *Iliad* 23. 257–897 (chariot race, boxing, wrestling, foot-race, fight in armour, shot put, archery, and spear-throw). The narration of each competition is structured by an underlying script, which encompasses the following elements: the prizes are set up; the challenge is announced; competitors come forward; preparations for the competition are made; the contest takes place; and prizes are collected.[21]

On a much more modest scale, we observe repetition in the paired dining scenes of *Odyssey* 1, when the meal Telemachos shares quietly with the goddess Athene is described (at 130–143), and is followed immediately by the account of the rowdy meal-time of the suitors (144–154). Finally, I draw attention to an extraordinary instance of serial repetition at *Odyssey* 14. 111–522. Here we observe a single speech routine (a speech format which I call 'the bid') used intensively—six times!—to shape a prolonged episode: the negotiations between Odysseus, disguised as a beggar, and his swineherd Eumaios.[22] Within this sequence Odysseus tries to do a deal with his swineherd, who responds to his bids in interesting ways. The script for a bid comprises two elements only: request and offer. If the terms are accepted (as will—eventually—happen), a third stage ensues: exchange.

(1) Odysseus proposes a bargain (14. 111–147): the confirmation of Eumaios' loyalty in return for information about Odysseus' whereabouts;
(2) Odysseus' second attempt (14. 148–173): a mantle and a cloak in return for news of Odysseus' whereabouts;

21 Minchin, *Homer and the Resources of Memory*, 39–45.
22 Elizabeth Minchin, "Serial Repetition in Homer and the 'Poetics of Talk': A Case Study from the *Odyssey*". *Oral Tradition* 14 (1999): 336–353.

(3) Eumaios proposes a bargain (14. 174–190): information about Telemachos' present situation in return for the beggar's own tale;
(4) Odysseus' third offer (14. 191–389): Odysseus offers his 'life story'; in return he wants his news of Odysseus' return to be accepted;
(5) Odysseus' fourth offer (14. 390–456): he requests a mantle and a tunic and a passage home (if Odysseus returns home) in return for his own life (if Odysseus does not);
(6) a final approach: Odysseus makes a bid for a cloak (14. 457–522); he offers a story to illustrate Odysseus' wily character in return for a warm cloak for the night.
This final bid is accepted and the exchange is completed.

What is significant about serial repetition? Repetition at the level of type-scene, like repetition of a formulaic phrase, offers the poet a ready solution to the challenges of composition in performance. It is an economical strategy that facilitates fluency. As for the members of his audience, they have some extra moments for evaluating the discourse (for paying attention to what is happening between the characters involved), by virtue of the reduced processing load that comes with repetition of this kind. Serial repetition may convey a contrast, as in the two meals described above at *Odyssey* 1. 125–154, and as in *Iliad* 23, where the different events of the funeral games allow us to encounter for the last time the cast of players on the Achaian side, each of whom behaves true to his character. Indeed, what is extraordinary about the long Odysseus-Eumaios scene of *Odyssey* 14 is the amount it can tell us about each participant's character and about his motivations.

But repetition in *Odyssey* 14 serves a further function. Tannen tells us that repetition in conversation not only has cognitive advantages; it also creates rapport. In this Homeric sequence the poet depicts the gradual development of what appears to be rapport between Eumaios and Odysseus: at first they seem to be communicating at cross-purposes; but, from the time Odysseus tells his 'life-story' at 14. 191–389, the poet depicts a growing sympathy between the hero and the cautious Eumaios. Actual rapport, however, can occur only between the poet and his audience. The audience members, recognizing the intensity of the negotiation and observing the poet's game of repetition, are completely engaged—and respond with amusement as the poet reveals the persistent nature of their hero.

Structural repetition in stories. We find structural repetition of another kind in stories told within each epic. In the 1940s Wilhelm van Otterlo drew attention to a particular pattern we often find in the ancient epics when the same wording or the same idea appears at the beginning and the end of a story or an episode within a story.[23] He coined the term "ring-composition" to describe this observable surface pattern. Van Otterlo claimed that in Homer within any one episode or story one might observe a number of narrative elements handled in the order A, B, C . . . and then rehandled in the reverse order (. . . C, B, A), thus creating a number of "rings". He claimed that the poet deliberately structured his *narrative* clauses to fulfil this ring-like structure. My understanding of ring-composition differs significantly from that of van Otterlo, as I shall demonstrate; but there is no doubt that we can find instances of repetition in the small-scale stories that are included in the epics. Let us consider some examples.

At *Iliad* 7. 132–133 Nestor begins one of his autobiographical tales with the phrase, "[I]f only . . . I were in my youth, as (I was) when . . ." (αἲ γὰρ Ζεῦ τε πάτερ καὶ Ἀθηναίη καὶ Ἄπολλον, / *ἡβῷμ'* ὡς ὅτ' . . .); and at 7. 157 he returns to that theme as his closing words, "[I]f only I were young now . . ." (εἴθ' ὣς *ἡβώοιμι*, . . .). And at *Iliad* 24. 602–619, when Achilles tells the story of Niobe and her children in order to persuade Priam to eat with him, he creates two rings of correspondence, each one marked by repetition: he introduces his tale at 602, saying, "[F]or even Niobe, she of the lovely tresses, remembered to eat" (καὶ γάρ τ' ἠΰκομος Νιόβη *ἐμνήσατο σίτου*) and he concludes, at 613, with the words, "[B]ut she remembered to eat when she was worn out with weeping" (ἣ δ' ἄρα *σίτου μνήσατ'* , ἐπεὶ κάμε δάκρυ χέουσα). Within the tale itself he creates a further ring, at 603 and 609, when he refers to the death of Niobe's children through the repetition of the verb: her twelve children were destroyed (603) (τῇ περ δώδεκα παῖδες ἐνὶ μεγάροισιν *ὄλοντο*), but the two (gods), though they were only two, destroyed all the others (609) (τὼ δ' ἄρα καὶ δοιώ περ ἐόντ' ἀπὸ πάντας *ὄλεσσαν*).

How might we interpret these repetitions? The formal features of Homer's narratives may be fruitfully compared with oral anecdotes told in the Western world today.[24] For an important study of contemporary anecdotes of this kind I refer to William Labov and Joshua Waletzky's investigation into the structure of stories.[25] Their study has shown that most oral anecdotes in our own western tradition reveal a common framework—a framework that we recreate automatically when we tell a story. This framework comprises a series of properties that

---

23 Willem van Otterlo, *De Ringkompositie als Opbouwprincipe in de Epische Gedichten von Homerus*. Amsterdam 1948.

24 Minchin, *Homer and the Resources of Memory*, 181–202.

25 William Labov and Joshua Waletzky, "Narrative Analysis: Oral Versions of Personal Experience". In: June Helm (ed.), *Essays on the Verbal and Visual Arts*, 12–44. Seattle 1967; William Labov, *Language in the Inner City. Studies in the Black English Vernacular*. Philadelphia 1972; Michael Bamberg (ed.), *Oral Versions of Personal Experience: Three Decades of Narrative Analysis*. Mahwah, NJ 1997.

have been identified as abstract, orientation, complicating action, evaluation, resolution, and coda. Labov elaborates on these designations through a series of questions: abstract (what is this about?); orientation (who? when? what? where?); complicating action (then what happens?); evaluation (so what?); resolution/result (what finally happens?); coda (that was that!).[26] Livia Polanyi subsequently proposed a further two categories, which frame the narrative itself: entrance talk, which precedes the abstract ("Have I told you about the time when ...?"); and exit talk, which bridges the gap between the events of the story and the present time ("So this is why ...").[27]

This framework, when it is fully exploited, opens the way to the possibility of repetition or near-repetition of words or of ideas at two levels. It is possible to find a close correspondence between Polanyi's entrance talk and exit talk (the elements that focus on the storyteller/audience relationship), and/or between abstract and resolution—or abstract and coda (the elements that introduce the narrative and that bring it to an end). In the everyday context such correspondences may be spontaneous. The correspondence of abstract with resolution tells us that the speaker has brought his or her tale to the conclusion we were led to expect. Likewise, the correspondence of exit talk with entrance talk indicates to the listener that the story has reached its end and that he or she is now being restored once more to the topic of conversation. These correspondences *may* be marked by verbal repetition. To my knowledge, the possibility that such correspondences may occur within this framework—and what they might tell us—has not been studied. For this reason I initiate a discussion now, setting out below some advantages of repetition, with regard to both production (for the storyteller) and comprehension (for the listener).

In the Homeric texts, in those cases where one character tells a story to others, we find precisely the same correspondences across the story framework as we find in oral anecdotes today: that is, we may find one or, at the most, two pairs of correspondence, often marked by verbal repetition. But, whereas the correspondences in everyday storytelling are more or less unpremeditated and non-uniform, I argue, on the basis of the frequent occurrence of verbal repetition in ancient epic at these fixed points in the framework, that oral epic practice represents a standardization, or stylization, of our everyday habits. Let us consider a further example from the *Odyssey*, in which we observe the creation through repetition of two corresponding rings.

---

26 Labov, *Language in the Inner City*, 370.

27 Livia Polanyi, "Literary Complexity in Everyday Storytelling". In: Deborah Tannen (ed.), *Spoken and Written Language: Exploring Orality and Literacy*, 155–170. Norwood, NJ 1982.

When Odysseus reminds Penelope of how he fashioned their marriage bed from the bole of an olive tree (*Odyssey* 23. 183–204) he begins (183–188) with entrance talk, of which the central idea is: "What man has *put my bed in another place*?" (τίς δέ μοι *ἄλλοσε θῆκε* λέχος;). Then follows his story, which commences with an abstract (188–189): "a great distinguishing mark, *a special feature*, is built into that bed. I made it myself" (μέγα *σῆμα* τέτυκται/ ἐν λέχει ἀσκητῷ· τὸ δ' ἐγὼ κάμον οὐδέ τις ἄλλος). He concludes his tale with a coda, at 202: "So I declare this *special feature* to you" (οὕτω τοι τόδε *σῆμα* πιφαύσκομαι); and he uses exit talk, at 202–204, to return to the topic under discussion: "I do not know now whether my bed is still in its place or if some man has cut underneath the stump of the olive and *put it [the bed] in another place*" (οὐδε τι οἶδα, ... *λέχος*, ἦέ τις ἤδη/ ἀνδρῶν *ἄλλοσε θῆκε* ...). For examples of similar storytelling practice observe the single "ring" in Eumaios' tale at *Odyssey* 14. 462–506 (entrance talk at 462–468; exit talk at 503–506) and in the suitors' tale at *Odyssey* 24. 121–190 (abstract 123–124; coda 186).

Repetition in the context of storytelling, like repetition in type-scenes, makes composition easier for the poet, as Tannen has suggested, by the very fact that the repeated unit (no matter how small) has already been used; and, as we have noted above, it makes listening easier for the audience. But repetition plays another important role in small-scale stories: the role of *delineation*, which serves the storyteller and his listeners in similar ways. I begin with the singer. When he first announces the end point of his story, whether in his entrance talk or in the abstract of his tale, he sets up that end point as a goal, the point towards which he must direct his story. It is not sufficient for the singer that he simply hold that goal in memory. Note that he actually verbalizes it; he "puts it out there" as an auditory cue that will prompt him as he tells his story. That is one function, a local function, of delineation. A second function relates to the storytelling task as a whole. We all know that when we are faced with a large and complex task it is more efficient to break that task into smaller well-defined units, and to concentrate our attention on one unit at a time. This in effect is how the oral epic poet composes. He traces a path through his narrative one task at a time, using the structural strategies of storytelling as one of his management tools. The poet in this epic tradition may not have been fully aware of the script-like story structure that he followed each time he told a story or narrated an episode from within a story. But he was sufficiently aware that, as he approached its conclusion and as he repeated key words from his abstract, he was tying off his narrative with pleasing complementarity and thus assuring his listeners that his earlier promise had been fulfilled.

For the audience members, on the other hand, the announcement, in advance, of the story's proposed end-point (through the abstract) likewise guides their expectations as they listen. The repetition of those introductory words or that initial idea, at the close of the tale, may represent the actual end-point of the narrative sequence (resolution); or it may seal off the story as a whole (coda). Repetition, therefore, delineates the small-scale story and measures out for the listener the progress of the larger tale.

Repetitions in similes. One of the defining features of oral epic is the extended simile, in which one idea or entity, such as a hero's performance in battle (a target domain), is compared with another idea or entity that has similar features (a source domain). Just as a simile in English is introduced by words such as "like" or "just as", in the ancient epics ὡς (just as) indicates that a simile is to follow. This introductory element will be followed by a verbal description of a vivid image that relates in crucial ways to the object of comparison—although, necessarily, in some way it also differs from it. It is not surprising that we prefer concreteness and "pictureability" in the source domain of similes—that is, we prefer material that can readily be stored in and accessed as an image from visual memory.[28] As we process the comparison, we map the material of this image over the target domain. Image and relevant cultural knowledge combine to assist us in 'reading' the simile and relating it to the narrative event that it elucidates.[29]

Let us turn to the ancient epics for some examples. When, for instance, the poet of the *Iliad* describes the prompt effect of the healing herbs that Paiëon applies to Ares' wound, he draws on a cheese-making simile: he compares the efficacy of the herbs that stem the flow of Ares' blood to the speed with which the juice of a cutting from a fig tree will turn liquid milk to curds. To follow the simile we must be able to envisage the process in our mind's eye and to recognize the wonder that accompanies such a transformation, although we may not understand its chemistry:

> ὡς δ' ὅτ' ὀπὸς γάλα λευκὸν ἐπειγόμενος συνέπηξεν
> ὑγρὸν ἐόν, μάλα δ' ὦκα περιτρέφεται κυκόωντι,
> ὥς ἄρα καρπαλίμως ἰήσατο θοῦρον Ἄρηα.
>
> As when the juice of the fig in white milk rapidly fixes
> that which was fluid before and curdles quickly for one who
> stirs it; in such speed as this he healed violent Ares.

Homer, *Iliad* 5. 902–904 (transl. Lattimore)

---

28 Allan Paivio, "The Mind's Eye in Arts and Science". *Poetics* 12 (1983), 1–18, 7–8.
29 Allan Paivio, *Imagery and Verbal Processes*. New York, 1971; Minchin, *Homer and the Resources of Memory*, 26–28, 132–160.

If storytellers were asked why they had decided to generate a simile at this or that point of their narrative, they would reply that the action of the tale had prompted a memory: a vivid image that appeared in their "mind's eye". Because this image caught the essence of the action in some way, the storyteller judged that the pictorial quality of the image would be helpful to his audience—and would enhance the reception of his tale. This image leads the storyteller in turn to the words that describe it—a prompting phenomenon that is described in Paivio's dual coding hypothesis.[30]

My reconstruction of this mental process leads me to ask *why* a simile might be effective at any point of a narrative. Andrew Goatly has proposed a useful catalogue of functions, which range across explanation and modelling, reconceptualization, the expression of emotional attitude, decoration, enhancing memorability, foregrounding, informativeness, the cultivation of intimacy, and the prolongation of the audience's pleasure.[31] All of these functions may be identified not only in everyday storytelling and literary discourse, but also in Homeric epic. The question of *how* these functions are served by a simile may be answered by considering the operations of the simile in terms of both parallelism and repetition. Since the image or scene at the heart of a simile is, as I noted above, mapped over a second (possibly less familiar) action or scene, the simile will lead the listener to a new, or richer, understanding of the target domain. Should its content be close to the audience's own experience, it will create a sense of intimacy with the storyteller. And, by virtue of parallelism and of the resultant delay as the action of the narrative is suspended, the simile draws attention to the event of the target domain and makes it memorable.

In Homeric epic we may find that repetition highlights an action or an emotion that lies at the heart of both the target domain and the source domain. As the poet moves from domain to domain he may reuse relevant words or phrases: an economical measure from the point of view of production. But such repetition also promotes comprehension and connection. Thus, at *Iliad* 13. 389–392 (and 16. 482–485), the verb of the simile repeats the verb of the narrative:

> *ἤριπε* δ' ὡς ὅτε τις δρῦς *ἤριπεν* ἢ ἀχερωῒς
> ἠὲ πίτυς βλωθρή, τήν τ' οὔρεσι τέκτονες ἄνδρες
> ἐξέταμον πελέκεσσι νεήκεσι νήϊον εἶναι· . . .
> ὣς ὁ πρόσθ' ἵππων καὶ δίφρου κεῖτο τανυσθείς.

---

30 Paivio, "The Mind's Eye"; Allan. Paivio and Mary Walsh, "Psychological Processes in Metaphor Comprehension and Memory". In: Andrew Ortony (ed.), *Metaphor and Thought*, 307–328, 2nd edn. Cambridge 1993; Alan Baddeley, *Human Memory: Theory and Practice*, 106–109. Hove and London 1990.

31 Andrew Goatly, *The Language of Metaphors*, 148–167. London and New York 1997.

> He *fell*, as when as oak tree *falls* or a white poplar
> or like a towering pine tree which in the mountains the carpenters
> have hewn down with their whetted axes to make a ship timber.
> So he lay there stretched out in front of his horses and chariot.
>
> Homer, *Iliad* 13. 389–392 (transl. Lattimore, modified)

And at 11. 556–557, at the end of the long simile (at 548–555) that describes a lion being forced back from its prey, we find the repetition of the idea of reluctant retreat:

> ἠῶθεν δ' *ἀπονόσφιν ἔβη τετιηότι θυμῷ*:
> ὣς Αἴας τότ' *ἀπὸ Τρώων τετιημένος ἦτορ*
> *ἤϊε* . . .
>
> and with the daylight *he goes away, disappointed of desire*;
> so *Aias, disappointed at heart, drew back* from the Trojans . . .
>
> Homer, *Iliad* 11. 556–557 (transl. Lattimore)

Note that, although the exact verb is not reused, the idea conveyed by the verb is the same (goes away/drew back); and the key descriptor "disappointed at heart" is rendered in almost exact terms. For a similar example see *Iliad* 2.475 (διακρίνωσιν, they separate) and 476 (διεκόσμεον, they divided, marshalled).

A distinctive feature of the Homeric simile is the standardized use (with few exceptions) of the correlatives ὡς . . . ὣς . . . (just as . . . so) to mark the poet's entry into the simile and, later, his return to his narrative. Just as the poet took care to delineate his small-scale stories and distinguish them from the narrative proper, so he keeps his similes separate from his narrative (Beck 2014). He maintains oversight of his performance, marking off the simile as a separate compositional task; in this way too he maintains connection with his listeners—keeping in mind their ability to follow his tale.

3 Conclusions

Deborah Tannen has identified four functions of repetition in spontaneous speech. These four functions—production, comprehension, connection, and interaction—contribute in varying degrees to the coherence of everyday talk. This is the case also in oral discourse in a formal context (I refer here specifically to oral traditional epic). I have shown in the examples above—of repetition and serial repetition in type-scenes, repetition within small-scale stories, and repetition and parallelism in similes—that the oral poet (unlike Anglophone speakers and writers) does not shy away from repeating himself in composition, since repetition will assist him in maintaining fluency: that is, in *production*. We observe that he recognizes, perhaps intuitively, that these same repetitions assist his audiences in their *comprehension* of his tales. Furthermore, the poet exploits the potential of repetition to serve the function of *connectedness*: not only to mark out the progression of his tale (as he defines the outer limits of a story or of a simile before he resumes the narrative proper) but also, in a subtle way, to tell his audience what is important (for example, through parallelism and repetition in his similes, or the repetition of a key event in a story). Finally, through repetition the poet *interacts* with his audience, demanding that his listeners share his pleasure in his tale—for example, as I noted above, his amusement at the behaviour of Odysseus at *Odyssey* 14. 111–522—as they process the same verbal format between the same characters again and again.

In oral storytelling of this formal kind repetition at a structural level is more intensive and, therefore, more predictable than it is in everyday discourse or, indeed, in literary discourse. I conclude that practitioners of this oral epic tradition that we associate with Homer identified over time a small range of narrative features—principally, type-scenes (including verbal scripts or formats), similes, and stories—and that, for the convenience of the poet, the comfort of the listener, and the effectiveness of the tale, the presentation of each 'genre' came to be standardized. This tradition has used and re-used those standardized forms, building into them further opportunities for repetition. Repetition, therefore, serves to conjure up a world in which nothing is hurried: actions are elaborated at a deliberate pace; even in the heat of battle words are exchanged with a degree of formality; similes may be used both to elucidate and to highlight; and through repetition similes, as well as inserted stories, are marked off from the narrative proper. Audience members not only tolerate repetition; they expect it at certain points and they depend on it; for them repetition is a defining aspect of oral song. This is the point at which production, comprehension, connectedness, and interaction coincide: the poet uses repetition in this high-pressure context as a device to manage the business of storytelling.

Tannen argues that the four functions that she has identified in everyday conversation lead to coherence through interpersonal *involvement* and the *rapport* of the participants: when one speaker repeats the words of another, for example, talk becomes a show of acceptance, of involvement, and of a willingness to interact.[32] We find illustrations of this same rapport when we turn to oral traditional epic. Homer almost unfailingly *depicts* interactive repetition in the

---

32 Deborah Tannen, "Repetition in Conversation". *Language* 63 (1987): 574–605; Deborah Tannen, "Repetition in Conversation as Spontaneous Formulaicity". *Text* 7 (1987): 215–243.

conversations he reports. For example, when a question is answered by another speaker, the respondent invariably repeats the content of the question: thus, at *Iliad* 6.376–380, Hektor quizzes the palace attendants about the whereabouts of Andromache and, at 382–389, the housekeeper responds using much the same language.

Finally, the poet achieves *coherence* through his deployment of familiar strategies—namely, standardization and repetition at all levels from repeated formulas and standardized type-scenes to the structural repetitions that I have discussed above. As I have shown in the case of serial repetition of type-scenes, the poet creates new meaning by using these same strategies in *novel* ways.[33] The combined effect of repetition and standardization is not, therefore, purely instrumental: standardization and repetition do not serve the functions of production and comprehension alone. This audience enjoys the poet's play on the patterns they have come to know; it is essential to their experience of oral song. They find pleasure in traditional forms and traditional strategies; and it gives them pleasure to observe the poet working inventively with this same material. Repetition and standardization are traditional devices, refinements of the casual practices of everyday conversation; but they may also be innovative and, in our terms, poetic.

References

Baddeley, Alan. *Human Memory: Theory and Practice.* Hove and London 1990.

Bakker, Egbert J. *Poetry in Speech: Orality and Homeric Discourse.* Ithaca and London 1997.

Bamberg, Michael (ed.). *Oral Versions of Personal Experience: Three Decades of Narrative Analysis.* Mahwah, NJ 1997.

Beck, Deborah. "Simile Structure in Homeric Epic and Vergil's Aeneid". In: Ruth Scodel (ed.), *Between Orality and Literacy: Communication and Adaptation in Antiquity. Orality and Literacy in the Ancient World X*, 244–266. Leiden 2014.

Biber, Douglas. *Variation across Speech and Writing.* Cambridge 1988.

Chafe, Wallace. "Integration and Involvement in Speaking, Writing, and Oral Literature". In: Deborah Tannen (ed.), *Spoken and Written Language: Exploring Orality and Literacy*, 35–53. Norwood, N.J. 1982.

Fahnestock, Jeanne. *Rhetorical Style: The Uses of Language in Persuasion.* Oxford 2011.

Friedrich, Paul. *The Language Parallax: Linguistic Relativism and Poetic Indeterminacy.* Austin 1986.

Goatly, Andrew. *The Language of Metaphors.* London and New York 1997.

Labov, William. *Language in the Inner City. Studies in the Black English Vernacular.* Philadelphia 1972.

33 Tannen, "Repetition in Conversation", 585.

Labov, Willaim and Joshua Waletzky. "Narrative Analysis: Oral Versions of Personal Experience". In: June Helm (ed.), *Essays on the Verbal and Visual Arts*, 12–44. Seattle 1967.
Lattimore, Richmond. *The* Iliad *of Homer*. Chicago 1951.
Lattimore, Richmond. *The* Odyssey *of Homer*. New York 1965.
Lord, Albert B. *The Singer of Tales*. Cambridge, MA 1960.
Minchin, Elizabeth. "Serial Repetition in Homer and the 'Poetics of Talk': A Case Study from the *Odyssey*". *Oral Tradition* 14 (1999): 336–353.
Minchin, Elizabeth. *Homer and the Resources of Memory: Some Applications of Cognitive Theory to the* Iliad *and the* Odyssey. Oxford 2001.
Minchin, Elizabeth. *Homeric Voices: Discourse, Memory, Gender*. Oxford 2007.
Minchin, Elizabeth. "Poet, Audience, Time, and Text: Reflections on Medium and Mode in Homer and Virgil". In: Ruth Scodel (ed.), *Between Orality and Literacy: Communication and Adaptation in Antiquity. Orality and Literacy in the Ancient World X*, 267–288. Leiden 2014.
Mueller, Martin. "Repetitions". In Margalit Finkelberg (ed.), *The Homer Encyclopedia*, vol. 3, 739–742. Chichester 2011.
Paivio, Allan. *Imagery and Verbal Processes*. New York 1971.
Paivio, Allan. "The Mind's Eye in Arts and Science". *Poetics* 12 (1983): 1–18.
Paivio, Allan. *Mental Representations: A Dual Coding Approach*. New York 1986.
Paivio, Allan and Mary Walsh. "Psychological Processes in Metaphor Comprehension and Memory". In: Andrew Ortony (ed.), *Metaphor and Thought*, 307–328. 2nd edn, Cambridge 1993.
Parry, Adam (ed.). *The Making of Homeric Verse: The Collected Papers of Milman Parry*. New York and Oxford 1971/1987.
Polanyi, Livia. "Literary Complexity in Everyday Storytelling" In: Deborah Tannen (ed.), *Spoken and Written Language: Exploring Orality and Literacy*, 155–170. Norwood, N.J. 1982.
Russo, Joseph. "Formula". In: Margalit Finkelberg (ed.), *The Homer Encyclopedia*, vol. 1, 296–298. Chichester 2011.
Russo, Joseph. "Sicilian Folktales, Cognitive Psychology, and Oral Theory". In: Thomas M. Falkner, Nancy Felson, and David Konstan (eds.), *Contextualizing Classics: Ideology, Performance, Dialogue*, 151–171. Lanham 1999.
Tannen, Deborah. "Oral and Literate Strategies in Spoken and Written Narratives". *Language* 58 (1982): 1–21.
Tannen, Deborah. "Repetition in Conversation: Towards a Poetics of Talk". *Language* 63 (1987): 574–605.
Tannen, Deborah. "Repetition in Conversation as Spontaneous Formulaicity". *Text* 7 (1987): 215–243.
van Otterlo, Willem. *De Ringkompositie als Opbouwprincipe in de Epische Gedichten von Homerus*. Amsterdam 1948.

Raymond F. Person, Jr., Ohio Northern University

From grammar in everyday conversation to special grammar in oral traditions: A case study of ring composition[1]

Abstract: Albert Lord described the oral traditional system of formulas and themes as a "special grammar" that was an extension of the grammar in conversation. Although Lord's description has been widely accepted, few studies have compared the special grammar of oral traditions to the grammar of everyday conversation. This chapter provides one case study—that is, an analysis of ring composition with examples taken from Serbo-Croatian epic, Homer, *Beowulf*, and Genesis and selected examples from conversation that are analogous to ring composition. This analysis demonstrates that even the most elaborate examples of ring composition can be understood as expansions of common practices found in everyday conversation. Thus, this linguistic analysis strongly suggests that the cognitive processes behind the special grammar of oral traditions are analogous to those of the grammar of everyday conversation.

1 Introduction

In his essay "Poetry of Grammar and Grammar of Poetry" Roman Jakobson wrote the following:[2]

> Let us insist on the strikingness of these [literary] devices; any sensitive reader, as Sapir would say, feels instinctively the poetic effect and the semantic load of these grammatical appliances, "without the slightest attempt at conscious analysis", and in many cases the poet himself in this respect is similar to such a reader. In the same way both the traditional listener and the performer of folk poetry, which is based on a nearly constant parallelism, catches the deviations without, however, being capable of analyzing them, as the Serbian guslars and their audience notice and often condemn any deviation from the syllabic patterns of the epic songs from the regular location of the break but do not know how to define such a slip.

---

1 This chapter is a revision of portions of Raymond F. Person, Jr., *From Conversation to Oral Tradition: A Simplest Systematics for Oral Traditions*. London, 2016. Permission has been granted for its use here by the commissioning editor of Routledge Press.

2 Roman Jakobson, *Selected Writings: III Poetry of Grammar and Grammar of Poetry*. The Hague 1981, 93.

Jakobson's argument assumes that the poetry of literature (like that of folk poetry) is somehow closely connected to language in general, so that the "sensitive reader" naturally understands the aesthetics without "conscious analysis". This assumption undergirds various approaches to language and literature and their relationship to cognition, including literary stylistics, discourse analysis, linguistic anthropology, historical pragmatics, and cognitive poetics.

The assumed close connection between literary devices and everyday language can be illustrated in the notion of 'foregrounding' in literary stylistics. In his handbook, Paul Simpson defined 'foregrounding' as follows:[3]

> *Foregrounding* refers to a form of textual patterning which is motivated specifically for literary-aesthetic purposes. Capable of working at any level of language, foregrounding typically involves a stylistic distortion of some sort, either through an aspect of the text which deviates from a linguistic norm or, alternatively, where an aspect of the text is brought to the fore through repetition or parallelism. That means that foregrounding comes in two main guises: foregrounding as 'deviation from a norm' and foregrounding as 'more of the same'.

Although Simpson's definition emphasizes "textual patterning", he nevertheless noted that foregrounding is "[c]apable of working at any level of language". In Jakobson's terms, foregrounding is a literary device based on deviations from non-literary language understood by the 'sensitive reader' without 'conscious analysis'. Foregrounding is one of the literary features that has been the object of significant empirical psycholinguistic study, so much so that in 2007 Willie van Peer could conclude as follows in his introduction to a special issue of *Language and Literature* devoted solely to foregrounding: "If the growth of empirical studies of foregrounding continues at the present rate, it may be expected that in a few decades the empirical evidence for the theory is going to look like that of a regular scientific theory".[4]

Jakobson referred to "the Serbian guslars and their audience", who have been the object of study by Milman Parry, Albert Lord, John Miles Foley, and others, especially as an empirical comparand to the study of literature with roots in oral tradition. This essay draws from the study of the comparative study of oral traditions, including Serbo-Croatian epic, as well as conversation analysis to show how the literary structure of ring composition can be understood as an example of what Simpson described as "foregrounding as 'more of the same'"—that is, ring composition is a literary device (found even in oral traditions) in

3 Paul Simpson, *Stylistics: A Resource Book for Students*. London 2004, 50.
4 Willie van Peer, "Introduction to Foregrounding". *Language and Literature* 16 (2007): 99.

which structures found in everyday conversation have been exaggerated for "literary-aesthetic purposes", so that a "sensitive reader" or an oral traditional audience can competently co-produce the traditional meaning.[5]

2 Ring composition: Definition and selective examples

In his study of ring composition in Homer, Steve Reece provided the standard definition in classics of the closely related hysteron-proteron and ring composition as follows:[6]

> By hysteron-proteron, I mean a pattern in which the last mentioned element of sequence becomes the first mentioned in the next (ABBA, ABCCBA, ABCDDCBA, etc.); ring composition is similar but is generally understood to include a central core (AXA, ABXBA, ABCXCBA, etc.).

Outside of classics (and sometimes within as well), these two structures are often treated under the same label of "ring composition" and this will be my approach here. In *Thinking in Circles*, Mary Douglas was primarily interested in the longer, more complex examples of ring composition. From her study, she identified the following rules or conventions, even though she noted that these are not hard and fast rules:[7]

1. *Exposition or prologue*: There is generally an introductory section that states the theme and introduces the main characters. ...
2. *Split into two halves*: ... The convention draws an imaginary line between the middle and the beginning, which divides the work into two halves, the first outgoing, the second returning. ...

5 This essay and the treatment of ring composition as a "case study" of the "special grammar" of oral traditions draws significantly from Raymond F. Person, Jr., *From Conversation to Oral Tradition: A Simplest Systematics for Oral Traditions*, London, 2016. There I analysis traditional phraseology (including formulas) and thematic structures as discussed by especially Lord and Foley from the perspective of conversation analysis to develop further the idea that the "special grammar" of oral traditions is based on adaptations of everyday language.

6 Steve Reece, "The Three Circuits of the Suitors: A Ring Composition in *Odyssey* 17–22". *Oral Tradition* 10 (1995): 211.

7 Mary Douglas, *Thinking in Circles: An Essay on Ring Composition*. Yale, 2007, 36–38.

3. *Parallel sections*: After the mid-turn the next challenge for the composer of the ring is to arrange the two sides in parallel ... by making separate sections that are placed in parallel across the central dividing line. ...
4. *Indicators to mark individual sections*: Some method for marking the consecutive units of structure is technically necessary. ...
5. *Central loading*: ... much of the rest of the structure depends on a well-marked turning point that should be unmistakable. ...
6. *Rings within rings*: ... the major ring may be internally structured by little rings.
7. *Closure at two levels*: By joining up with the beginning, the ending unequivocally signals completion. ... The final section signals its arrival at the end by using some conspicuous key words from the exposition.

Thus, Douglas explicated further what Reece summarized in the notation of ABCCBA and ABCXCBA.

Ring composition has been identified in a variety of literatures, especially ancient and medieval literature. Below I provide some representative examples from a variety of literatures, the first from Foley's study of the Christian Serbo-Croatian epic tradition—that is, an example from a living oral tradition.[8] The following are the opening lines from "Marko Drank Wine During Ramazan", which also occur in two other locations in this song with slight variations:[9]

Tsar Sulejman issued an order:
A that none drink wine during Ramazan,
B that none wear green dolamas,
C that none strap on plated swords,
D that none dance the kolo with women.
D Marko danced the kolo with women,
C Marko strapped on a plated sword,
B Marko wore a green dolama,
A Marko drank wine during Ramazan.

Here we clearly have an example of ring composition (ABCDDCBA) from an oral tradition.

8 See also Lord's discussion of ring composition in the Muslim Serbo-Croatian epic tradition. Albert Lord, "The Merging of Two Worlds: Oral and Written Poetry as Carriers of Ancient Values". In: John Miles Foley (ed.), *Oral Tradition in Literature: Interpretation in Context*, 53–64. Columbia 1986.

9 John Miles Foley, "Old English and Serbian Poetry". *Anglo-Saxon England* 12 (1983): 195. Translation Foley's.

Reece provided the following diagram summary of the series of questions that Odysseus asked his dead mother and her corresponding answers in reverse order (*Odyssey* 11.170–203), which form an extended ring composition:[10]

A – What killed you? (171)
B – A long sickness? (172)
C – Or Artemis with her arrows? (172–73)
D – How is my father? (174)
E – How is my son? (174)
F – Are my possessions safe? (175–76)
G – Has my wife been faithful? (177–79)

G – Your wife has been faithful. (181–83)
F – Your possessions are safe. (184)
E – Your son is thriving. (184–87)
D – Your father is alive but in poor condition. (187–96)
C – Artemis did not kill me with her arrows. (198–99)
B – Nor did a sickness kill me. (200–01)
A – But my longing for you killed me. (202–03)

Reece also noted comparative data for ring structures, including other literature with their roots in oral traditions (Old English, the Bible, Old French) as well as living oral traditions (African epic, Scottish ballad, and Serbo-Croatian epic).[11] On the basis of the comparative data, he concluded that ring composition is "pervasive in orally composed narrative".[12]

John Niles has described the entire structure of *Beowulf* as a "large-scale symmetry" and ring composition: "(*A*) Introduction, (*B*) fight with Grendel, (*C*) celebrations, (*D*) fight with Grendel's mother, (*C*) celebrations, (*B*) fight with dragon, (*A*) close".[13] He then demonstrated that each of the fight scenes consists of a ring composition as is illustrated in this chart of the fight with Grendel's mother:[14]

A Hrothgar's speech on Æschere's death and the description of the monster's pool (1321–82)
B Beowulf wants to avenge Æschere's death (1383–96)
C Danes and Geats go to the pool (1399–1421)

10 Reece, "The Three Circuits of the Suitors", 213–214. For her analysis of similar examples of ring composition composed of questions and answers, see Minchin, *Homeric Voices*, 102–116.
11 Reece, "The Three Circuits of the Suitors", 213–214.
12 Reece, "The Three Circuits of the Suitors", 207.
13 John D. Niles, *Beowulf: The Poem and Its Tradition*. Cambridge 1983, 157–158.
14 This chart is a summary of Niles' discussion of this passage. See Niles, *Beowulf*, 154–156.

D arming of Beowulf in his helm and byrnie (1441b–54)
X the battle and defeat of Grendel's mother (1455–1628)
D disarming of Beowulf in his helm and byrnie (1629–30a)
C Danes and Geats return to Heorot (1632–50)
B Beowulf reports his successful revenge (1651–76)
A Hrothgar's speech on pride (1700–84)

Thus, Niles concluded that "the poem consists of three major episodes of different length and complexity, each one of which shows ring patterning".[15]

Although she discussed other examples, Douglas devoted most of her attention to the Bible and Homer. Here I provide one of her examples, her analysis of Genesis 22 as summarized in the following chart:[16]

v. 1: God called "Abraham!" "Here I am"
A v. 2: "Take your son, your only son"
B v. 3–6: Going to the place. Abraham saw the place afar off.
C v. 7–8: "Where is the lamb for the burnt offering?" "God will see to it, my son." So they went on together.
D v.9–10: Abraham raised his hand, took the knife to slay his son.
X v. 11: The Angel of the Lord: "Abraham, Abraham!" "Here I am."
D v. 12: "Do not raise your hand against the lad"
C v. 13: Abraham saw the ram, took it. Offered it as a burnt offering.
B v. 14: Called the name of the place, "God will provide"
A v. 15–18: "You have not withheld your son, your only son."

Thus, we have seen examples of ring composition from a living oral tradition—Serbo-Croatian—and from three different literary traditions that have been understood by some scholars as being influenced by the aesthetics of oral traditions—Homer, *Beowulf*, and the Bible.

3 Ring composition: The controversy concerning cognition

Early critics of Parry and Lord complained that their understanding of the compositional process was based on rules that were far too rigid and mechanical, thereby denying Homer's creativity. Often behind these criticisms was also an assump-

15 Niles, *Beowulf*, 157.

16 Douglas, *Thinking in Circles*, 20, 23. This chart combines her analysis in these two charts and reformats them for ease of reproducing.

tion that such literary genius as exhibited by Homer could not possibly come from the messy world of oral discourse but most certainly required writing—that is, great literature requires writing as its medium, because oral composition is necessarily too inexact to produce masterpieces like the *Iliad*. Although this line of criticism is based on misunderstandings of Parry and Lord's work,[17] it nevertheless influenced some subsequent assessments of the Parry-Lord approach, probably because the criticism is based on cultural assumptions commonly shared among modern Western academics.

What this disagreement illustrates is a fundamental difference of opinion concerning the cognitive-linguistic resources available without the aid of the technology of writing. The critics of oral poetics often assume that the technology of writing dramatically changed human cognition so that only written literature can have the complex literary structures that are the result of and evidence of the creativity of literary masters. Thus, in their opinion, oral poets lack the cognitive-linguistic resources to produce literature or at least the same quality of literature. The oralists' response against such arguments began with Lord and consists of two strategies. First, comparative evidence from living oral traditions collected by ethnographic research provides counter-examples—that is, examples of various so-called "literary" structures produced by oral poets.[18] Second, Lord described the oral compositional system as being analogous to language in general—that is, a system with grammatical rules and structures that nevertheless allows for originality and creativity:[19]

> The classical grammar of a language, with its paradigms of tenses and declensions, might give us the idea that language is a mechanical process. The parallel, of course, goes even further. The method of language is like that of oral poetry, substitution in the framework of the grammar. Without the metrical restrictions of the verse, language substitutes one subject for another in the nominative case, keeping the same verb; or keeping the same noun, it substitutes one verb for another. In studying the patterns and systems of oral narrative verse, we are in reality observing the "grammar" of the poetry, a grammar superimposed, as it were, on the grammar of the language concerned. Or, to alter the image, we find a special grammar within the grammar of the language, necessitated by the versification. The formulas are the phrases and clauses and sentences of this specialized poetic grammar. The speaker of this language, once he has mastered it, does not move any more mechanically within it than we do in ordinary speech. When we speak a language, our native language, we do not repeat words and phrases that we have memorized consciously, but the words

---

17 See especially Lord's own response to these early critics in Albert Lord, "Homer's Originality: Oral Dictated Texts". *Transactions of the American Philological Association* 84 (1953): 124–134.
18 For excellent introductions to various living oral traditions, see John Miles Foley, *Teaching Oral Traditions*. New York 1998 and the synopses in the special volume of *Oral Tradition* 18 (2003).
19 Albert Lord, *Singer of Tales*. Cambridge 1960, 35–36.

> and sentences emerge from habitual usage. This is true of the singer of tales working in his specialized grammar.

Lord's assertion that the language of oral traditions is a special grammar has been widely accepted by scholars in the Parry-Lord school. For example, Egbert Bakker observed as follows:[20]

> poetic meter and formulas, rather than removing Homeric poetry from the realm of the ordinary and the everyday, derive from what is most natural in spoken discourse: the "chunks" that make up the adding style. I argue that meter and formulas entail the stylization of ordinary speech, rather than some inherently poetic principle.

Elizabeth Minchin concluded that "the *Iliad* and the *Odyssey* are instances of 'special' storytelling" and that "Homer's mimesis of speech acts is an echo of everyday discourse from his own world".[21] The most thorough adaptation and extension of Lord's notion of special grammar is in the work of John Miles Foley, who concluded as follows:[22]

> And what do we make of the result—a complex, heterogeneous, ever-evolving collection of inequivalent elements overseen by rules and processes no singer ever consciously imagined? Briefly stated, what we gain, apart from a philologically sound profile, is a foundation for aesthetic inquiry that is firm because it is faithful to the language and poetics of Serbo-Croatian. By formulating rules for the phraseological events we perceived as lines, we begin to restore a lost complexity to oral traditional diction; in effect, the point of view advocated in this chapter, and for that matter throughout the volume, allows us to "re-complicate" poetic composition, to take it out of the arena of lockstep simplicity and back to the realm of language—the most complex of human abilities and arts.

Some of the early critics of Parry and Lord included discussions of ring composition in their critiques. For example, Calvin Kendall concluded concerning *Beowulf* that "the style and structure of the poem seem thoroughly literary".[23] He supported this contention by referring to the work of John Niles on the use of ring composition as a structural device in *Beowulf*. In Kendall's opinion, Niles had convincingly argued that ring composition as found in the *Iliad* and *Beowulf* has a complexity lacking in those few examples found in oral traditions obtained in

---

20 Egbert Bakker, *Poetry in Speech: Orality and Homeric Discourse*. Ithaca 1997, 2.

21 Elizabeth Minchin, *Homer and the Resources of Memory: Some Applications of Cognitive Theory to the* Iliad *and the* Odyssey. Oxford 2001, 203; Elizabeth Minchin, *Homeric Voices: Discourse, Memory, Gender*. Oxford 2007, 19.

22 John Miles Foley, *Traditional Oral Epic: The Odyssey, Beowulf, and the Serbo-Croatian Return Song*. Berkeley 1990, 199–200.

23 Calvin B. Kendall, *The Metrical Grammar of Beowulf*. Cambridge 1991, 6.

fieldwork.[24] Niles may well have been correct that at that time more field work needed to be done, but his wording is actually more tentative than Kendall's conclusion: "[t]he various sorts of ring composition in *Beowulf* may plausibly be taken as traits that would be useful to an oral poet or performer, but further conclusions are premature".[25] However, Niles nevertheless sets up a strong contrast between orally-derived works and literature.[26]

> Far from having been an unskilled compiler of separate tales, the poet was endowed with a keen (although always flexible) sense of narrative form. ... *Beowulf* is no mere collection of "fabulous exploits redolent of folk-tale fancy." It is no sightless narrative, nor is it a clumsy joining of two tales. It is a well-wrought epic poem.

Orally-derived works seem to be characterized as a "mere collection of 'fabulous exploits redolent of folk-tale fancy'" coming from "*unskilled* compilers of separate tales" producing a "*sightless* narrative" or a "*clumsy* joining of two tales." In contrast, literature is produced by a poet "with a *keen* ... sense of narrative form" producing a "*well-wrought* epic poem." Thus, both Kendall and Niles assumed the "literary" complexity of ring composition, at least that of the "quality" found in "great" works of literature, to argue that these same "great" works are likely not the result of the compilation of oral compositions, which is characterized by descriptions such as "unskilled compilers" producing inferior narratives lacking literary unity.

A similar understanding is evident in Douglas's study of ring composition. Noting how widespread it is in world literature, she described ring composition as "a worldwide method of writing".[27] Referring to the work of Lord, she wrote, "Among speculations about the origin of this complicated rhetorical system, a common one is that it comes from pre-literate times where it was a necessary help to the memory of the bards".[28] However, she seems to have quickly dismissed such "speculation" when she wrote, "Whatever its pre-literate origins, ring composition has of course been transformed with the advent of writing".[29] Since most of her own examples are from the Hebrew Bible, she understood that most examples of ring composition come from "literate" cultures. "The prolific biblical

---

24 Kendall quoted Niles's conclusions from the end of his chapter 7 entitled "Ring Composition" (*Beowulf*, 160), from which the following quotes are also taken.
25 Ibid.
26 Ibid.
27 Douglas, *Thinking in Circles*, x.
28 Ibid, 12.
29 Ibid.

examples of parallelism come from a very literate civilization".[30] In fact, she drew from the work of Jack Goody to insist that "learning by rote is a characteristic of literate society" and ring composition was a mimetic aid for such memorization.[31]

As we can see in Kendall's and Douglas's remarks, even evidence of ring composition in living oral traditions, such as Lord's and Foley's study of Serbo-Croatian, has not convinced some of how ring composition can be a sophisticated structure employed by oral poets. This strongly suggests an assumption that oral discourse is simply too messy to give rise to such "literary" structures as ring composition. This oppositional thinking can be summarized in the following binary oppositions: oral versus literate, illiterate versus literate, simple versus complex, mechanistic versus creative, etc. Such binary thinking has been described as the "Great Divide" thesis and is widely rejected by scholars of oral poetics. Nevertheless, the influence of the "Great Divide" thesis continues. Thus, the second strategy of demonstrating how ring composition belongs to the special grammar of oral traditions that is adapted from practices in everyday conversation remains helpful.

Lord, Bakker, and Minchin have written on ring composition as belonging to the special grammar of oral traditions. Lord concluded his discussion of ring composition by drawing from the earlier work of Samuel Basset on Homer and Fred Robinson on *Beowulf*:[32]

> We should be grateful to Bassett for his perspicacity in pointing out the obvious, that Homer *preceded* the rhetoricians, that the "Homeric" hysterion proteron and ring composition were in the ancient Greek oral-traditional style before the rhetoricians labeled them "chiasmus." They were not the inventions of the rhetoricians, but if Bassett, taking his cue from Aristarchus, is correct, [they are] a natural order of continuous thought. ... A half-century after Bassett, Fred C. Robinson speaks of "calculated ambiguity" when discussing the boar's head image in *Beowulf* and the contribution of the appositive style to the appreciation of a possible layering of connotations. If I understand him alright, he points to the possible survival in the apposed elements of older meanings and references merged with the contemporary. Near the end of his last chapter, however, he prefers to think of apposition as

30 Ibid. Recently various biblical scholars have made arguments for ancient Israel as a primarily oral society, in which even written texts operated primarily in oral contexts. See Susan Niditch, *Oral World and Written Word: Ancient Israelite Literature*. Louisville 1996; David M. Carr, *Writing on the Tablet of the Heart: Origins of Scripture and Literature*. Oxford 2005; and Raymond F. Person, Jr., *The Deuteronomic History and the Book of Chronicles: Scribal Works in an Oral World*. Atlanta 2010.

31 Douglas, *Thinking in Circles*, 13.

32 Albert Lord, "Ring Composition in *Mladon*; or, a Possible Case of Chiasmus in a Late Anglo-Saxon Poem". In: , Joseph Harris (ed.), *The Ballad and Oral Literature*. Cambridge 1991, 242.

> a "habit of mind"—a phrase, it seems to me, that hits the mark very well—rather than as a rhetorical device.

Here we can see Lord's understanding of ring composition as an extension of his understanding of theme. If theme is a "group of ideas",[33] then ring composition is one particular structure by which groups of ideas may be organized. If "theme, even though it be verbal, is not any fixed set of words, but a grouping of ideas",[34] then ring composition is not "a rhetorical device" invented by the rhetoricians but "a natural order of continuous thought" or a "habit of mind" that underlies everyday language and the special grammar of oral traditions.

Bakker extended Lord's notion of a special grammar and made a similar observation concerning ring composition.[35]

> Ring composition, then, is less a feature of archaic style as such (in opposition to other styles) than an index of the ways in which this style, as special speech, draws on the resources of ordinary speech. Yet the formalization is never so total that the ring becomes a structure for its own sake in which the end is a return to the beginning, a structure that finds its fulfillment in a symmetrical array on paper. The term "ring," in fact, might invoke the wrong image, in that speech, which proceeds through time, does not and cannot go back to an earlier point; it has to move forward and in so doing cannot but be uttered within the context of previous speech and provide context for speech to come. The principle that repetition in speech is impossible insofar as no two contexts are exactly identical appears to apply quite strongly in the case of ring composition: between the indication of the goal and its achievement, the world has changed.

Thus, according to Bakker, ring composition is an adaptation or formalization of conversational structures that nevertheless allow for some flexibility within the structure itself.

Explicitly building upon the work of Bakker, Minchin concluded her chapter on *hysteron proteron* as follows:[36]

> My conclusions, therefore, are two: first, that this familiar practice, of answering questions in an order which reverses the sequence of asking, is motivated by cognitive factors (to make effective use of the resources of memory at our disposal) and social factors (co-operativeness above all); and, second, since *hysteron proteron* is observable not only in the conversations in which we all take part but also in Homeric discourse, that poets in this particular epic tradition have recognized this feature of oral discourse and, in mimicking it, have exploited its almost rhythmical regularity in their composition of oral song.

---

33 Lord, *Singer of Tales*, 68.

34 Ibid, 69.

35 Bakker, *Poetry in Speech*, 120–121.

36 Minchin, *Homeric Voices*, 115–116.

Lord's and Bakker's assertions that ring composition is an element of the special grammar of oral traditions were primarily intuitive observations lacking empirical support in the form of a clear explication of what resources in ordinary speech are the basis for this adaptation. In contrast, Minchin's analysis explicitly referred to observations from conversation analysis, including the basic observation of adjacency pairs.[37] However, Minchin's use of conversation analysis remains limited and, therefore, requires further elaboration.

4 Adjacency pairs and expansion

One of the basic observations of conversation analysis is sequence organization. As described by Emanuel Schegloff, "sequences of turns are not haphazard but have a shape or structure, and can be tracked from where they come from, what is being done through them, and where they might be going".[38] The most basic structure in sequence organization is adjacency pairs as characterized by Schegloff as follows:[39]

> In its minimal, basic unexpanded form an adjacency pair is characterized by certain features. It is:
> (a) composed of two turns
> (b) by different speakers
> (c) adjacently placed; that is, one after the other
> (d) these two turns are relatively ordered; that is, they are differentiated into "first pair parts" (FPPs, or Fs for short) and "second pair parts" (SPPs, or Ss for short). First pair parts are utterance types such as question, request, offer, invitation, announcement, etc.—types which *initiate* some exchange. Second pair parts are utterance types such as answer, grant, reject, accept, decline, agree/disagree, acknowledge, etc.—types which are *responsive* to the action of the prior turn.... Besides being differentiated into Fs and Ss, the components of an adjacency pair are
> (e) pair-type related; that is, not every second pair part can properly follow any first pair part. Adjacency pairs compose pair *types*; types are exchanges such as greeting-greeting, question-answer, offer-accept/decline, and the like. To compose an adjacency pair, the FPP and SPP come from the same pair type.

37 Ibid, 107–110.

38 Emanuel Schegloff, *Sequence Organization: A Primer in Conversation Analysis 1*. Cambridge 2007, 3. This section on sequence organization draws significantly from Schegloff's work as an excellent summary of the field.

39 Ibid, 13.

Minimal, two-turn adjacency pairs are common in opening and closing sequences in telephone calls. The following is from a closing sequence with two minimal adjacency pairs:[40]

Ile: Ta:ke keyuh
Cha: Speak tih you [(   )
Ile: [Bye: bye
Cha: Bye,

Because of this basic schema, expansions to this basic form can be found in three locations related to the two pair parts:[41]

<– Pre-expansion
First pair part
<– Insert expansion
Second pair part
<– Post-expansion

Pre-expansions can be in the form of pre-invitations, pre-offers, and pre-announcements—for example, "Didju hear the terrible news." is a pre-announcement before a storytelling.[42] Insert expansions can be in the form of clarifying questions that either seek more information concerning the first pair part or more information necessary before the second pair part can be provided. Post-expansions can be in the form of brief responses (for example, "oh" or "good") or more elaborate, such as disagreeing with the second pair part. The following example illustrates an adjacency pair with both an insert-expansion and a post-expansion.[43]

Sta: That's all. But you know what happened that night
we went to camp. Forget it. She wouldn't behave for
anything.
Ala: W-when.
Sta: When we went to camp.
Ala: She behaved okay.
Sta: She did?
Ala: Yeah. She could've been a lot worse.

Lines 2–3 contain an assessment ("wouldn't behave") that the other speaker will later disagree with. In line 4, the other speaker initiates a repair ("W-when")

40 Ibid, 23. Schegloff devoted an entire chapter to minimal, two-turn adjacency pairs.
41 Ibid, 26.
42 Ibid, 40.
43 Ibid, 154.

rather than agreeing with the assessment, thereby initiating an insert expansion. The first speaker completes the insert expansion in line 5 with the answer ("When we went to camp"). Now the second speaker disagrees with the initial assessment in line 6 ("behaved okay"). The first speaker then reasserts his initial assessment by questioning the disagreement in line 7 ("She did?"). Note that the assessment is now mitigated in that it occurs in the form of a question rather than a reassertion of the assessment (something like, "Oh no, she misbehaved"). This mitigation can be seen as reworking the initial assessment to some degree. Then in line 8 the other speaker begins with a confirmation of his disagreement ("Yeah"), but then mitigates his disagreement as well ("She could've been a lot worse"), thereby reworking the initial disagreement.

Here I have provided an example of an adjacency pair with a relatively brief insert expansion and a short post-expansion sequence; however, expansions can be quite elaborate and lengthy. For instance, Schegloff provided an example from a telephone call between a teenage couple consisting of 96 lines, in which the first pair part of the base adjacency pair is given in line 9 following a pre-expansion but the second pair part does not occur until line 93 followed by a post-expansion —that is, the entire 96 lines can be understood as one sequence built upon the base adjacency pair of request-acceptance with both a pre-expansion and post-expansion as well as an elaborate insert expansion. Thus, Schegloff can conclude as follows:[44]

> Expansions in each of these positions can be substantial, and (with a few exceptions) expansion can occur in all of them for any given sequence. As a result, then, very long stretches of talk can be understood as elaborate structures built around a single underlying adjacency pair.

Thus, within conversation basic structures that may typically be minimal, including adjacency pairs, can nevertheless be expanded significantly within the context of face-to-face talk-in-interaction.

5 Examples from conversation analogous to ring composition

Here I discuss three different sequences from naturally occurring conversation that have analogous structures to ring composition. Each of these examples

44 Ibid, 26–27.

includes expansions of the minimal basic adjacency pair. The first two examples can be understood as expansions built on one adjacency pair. The third example consists of an example in which the speaker produces a multi-unit turn consisting of three questions, thereby initiating three possible adjacency pairs. However, all three of these examples can be understood as expansions of adjacency pairs that are analogous to ring composition with the structure ABCCBA.

In the following example we can see how reported speech in the context of conversational storytelling can be built on the basis of an adjacency pair with post-expansion and can form what is analogous to the structure of ring composition found in the literary examples above.[45]

Prologue:	When we moved to Pennsylvania, Delbert and Earl walked to school by some neighbors.	
	And I met that lady one day when we were very new.	
A	and she said, "oh, you're the one with the twins."	[First pair part: Assessment]
B	and I said, "oh no, maybe you mean my boys that are a year and a half apart."	[Second pair part: Disagreement]
C	"oh no, they're twins." [laughing]	[Post-expansion: reworking first]
C	this lady was telling me, "oh no, they're twins."	
B	I said, "I have sons a year and a half apart."	[Post-expansion: reworking second]
A	"ah, well I think they look like twins."	[Post-expansion: reworking first]
Epilogue:	and I could've just throttled that woman=	
	I could've just *kicked* that woman.	
C	"oh no, they're twins." [laughing]	

Here a woman is telling a story about a conversation she had with a neighbor. Her narration of this reported speech forms what is analogous to an ABCCBA ring composition. This story consists of one base adjacency pair containing an assessment by the neighbor that the mother has twin sons and the mother's disagreement. The mother provides some background information in what I have labeled the prologue. She then reports what the neighbor said, "oh, you're the one with the twins". Note that this assessment assumes the validity of a prior assessment that the two boys are twins and is really a new assessment that the storyteller is the twins' mother. The mother's disagreement with the assessment actually addresses the prior assumed assessment that the boys are twins: "oh no, maybe you mean my boys that are a year and a half apart". That is, she is the mother of two boys who are not twins that *maybe* the neighbor is referring to. Note that the mother's dispreferred second is characteristically mitigated here with the use of

45 Neal R. Norrick, *Conversational Narrative: Storytelling in Everyday Talk*. Amsterdam 2000, 38.

"maybe you mean". The neighbor then asserts "oh no, they're twins", which is a form of post-expansion—that is, expansion after the completion of the second pair part of the base adjacency pair. In this case, she is reworking the first pair part to make clear that she is certain that the two boys she is talking about are twins, implying that maybe the mother is not the mother of the two boys she is talking about. The mother then repeats this reported speech in a way that emphasizes how absurd the conversation is from her perspective. The mother then reports her more assertive disagreement, lacking any hedges: "I have sons a year and a half apart". She then reports how the neighbor reworks her first pair part yet again: "ah, well I think they look like twins". With this example of ring composition from a story found in everyday conversation we can also see how the middle core can often be interpreted as the most important. In this case, what I have labeled as the epilogue includes the mother repeating the absurd thing the neighbor said—that is, the middle element of the ring composition.

The next two examples come from institutional settings. The first comes from an Oral Proficiency Interview of an employee of a Japanese company (C = Candidate) by a native-speaker of English (I= Interviewer) to assess the employee's advancement in the company's English language program.[46]

I:	How long can you live in the dormitory (0.3)	[First pair part: Question 1]
C:	I don't know about this dormitory but I thi::nk uh::: (0.2) I will stay (0.5) I will live there sec- um three or four years.	[Second pair part: Answer 1]
I:	uh-huh but how long can you live in the dormitory, What's the longest time you could live there (.) if you wanted to.	[Repair initiator: Reworking the first pair part: Question 1 & 2]
	(0.2)	[Pause indicating no answer]
I:	Forever?	[Repair initiator: Reworking the first part part: Question 3]
C:	No. Uh:::: se-seven years	[Answer to Q3; Answer to Q1–2]
I:	Okay. Seven years is the limit. Okay.	

The Candidate misunderstood the Interviewer's initial question; therefore, the Interviewer produces a turn with multiple questions, first repeating the question with a different prosodic emphasis ("uh-huh but how long can you live in the dormitory") and then rewording the question ("What's the longest time you could live there (.) if you wanted to."). Note that in the reworded question the pause (.)

46 Gabriele Kasper and Steven J. Ross, "Multiple Questions in Oral Proficiency Interviews". *Journal of Pragmatics* 39 (2007): 2057.

can be interpreted as space for the Candidate to respond before "if you wanted to" is added to the turn for prosodic elaboration. A longer pause follows (0.2) before the Interviewer now provides a possible answer to the initial question in the form of yet a third question ("Forever?"). The Candidate now understands the questions and then answers the last question first ("No." [not forever implied]) before answering the initial question ("seven years"). The Interviewer confirms receipt of the correct answer ("Okay. Seven years is the limit. Okay."). The series of three questions by the Interviewer can be represented as ABC within a multi-unit turn-at-talk and then the two answers within a multi-unit turn by the Candidate as CBA, with the understanding that "seven years" answers both of the first two questions. Therefore, this sequence is somewhat analogous to an ABCCBA ring composition.

The following example is similar to the previous one in that it is also from institutional talk and consists of a series questions in a multi-unit turn followed by corresponding answers in a multi-unit turn. This example comes from a White House press conference with President Ronald Reagan (=RR), but it illustrates a common device used by journalists (=JRN) once they have been called on—that is, since they are typically called on only once, their turn-at-talk often contains multiple questions on the same topic.[47]

JRN: Mr. President, who will be leading the U.S. delegation [Question 1]
to Leonid Brezhnev's funeral?
If you won't be going, how come? [Question 2]
And also, aside from your personal hopes for peace,
do you have reason to believe that the next coming months [Question 3]
might see the new Soviet leadership flexing its muscle a bit
and a period of increased tension coming about?
RR: Well, answering the last question first, no, I don't [Answer to Q3]
anticipate that as they make this transition. ...
But with regard to the service, we've had no direct, [Answer to Q1-2]
official word yet on anything about the service, although
we are in communication directly with them. And it was
just a plain case of looking at scheduled and my own schedule ...

The Journalist asks a series of three questions concerning the death of Soviet leader Brezhnev and American foreign policy and Reagan answers the questions in reverse order, even explicitly stating so ("answering the last question first").

47 Steven E. Clayman and John Heritage, "Questioning Presidents: Journalistic Deference and Adversarialness in the Press Conferences of U.S. Presidents Eisenhower and Reagan". *Journal of Communication* 52 (2002): 754.

Once again the first two questions are similar enough that one answer provides the answer to both questions. Therefore, this sequence is also analogous to an ABCCBA ring composition. Although this example has similarities to the previous example, it nevertheless differs in that the previous example can be understood as a sequence consisting of one base adjacency pair (question-answer) whereas this example clearly has three adjacency pairs of question-answer.

6 Conclusion

Although I have provided empirical examples from conversation that are analogous to ring composition and have suggested how ring composition in oral traditions and literature with roots in oral tradition may be explained without the need of literacy, I can imagine that some skeptics may still protest that my conversational examples are not nearly as elaborate as the examples from Homer, *Beowulf*, and Genesis. For example, the conversational ring compositions of ABCCBA do not compare to Homer's ABCDEFGGFEDCBA. At least this is exactly the type of argument that Kendall and Niles made concerning ring composition in *Beowulf* as compared to examples from field studies of living oral traditions. However, in *From Conversation to Oral Tradition*, I address this possible criticism in relationship to a variety of structures and their adaptations, including traditional phraseology and thematic structures, by noting how the communicative economy changes dramatically from face-to-face talk-in-interaction to storytelling in conversation and then even more so in oral traditions, thereby allowing expansion and exaggeration.[48] In face-to-face conversation we are constantly having to negotiate who gets the next turn at talk; therefore, turn construction requires close attention to turn allocation, in order to project the end of a turn-construction unit and possibly to select the next speaker. However, when a participant in a conversation wants to tell a story, the potential storyteller must negotiate with the other participants the ability to produce such a multi-turn sequence, thereby reaching an agreement to abandon the standard turn-taking system for the purpose of the storytelling. As soon as the story ends, the participants then return to the standard turn-taking system, once again constantly negotiating who gets the next turn at talk. Therefore, within storytelling in conversation the communicative economy changes by agreement of all of the participants who suspend

48 For my discussion of traditional phraseology and thematic structures as adaptations of conversational practices, see Person, *From Conversation to Oral Tradition*, Chapters 3–4.

the standard turn-taking system for the sake of the storytelling by making turn allocation less important.[49]

Oral traditions are a further extension of storytelling in conversation and the communicative economy is changed even more dramatically, in that turn allocation is suspended even more so for the sake of the (re-)performance. Therefore, if sequences in conversational storytelling can be analogous to an ABCCBA ring composition, we can explain how storytelling in the more institutional setting of oral traditional (re-)performance may contain more elaborate ring compositions, because of the change in the communicative economy allows for exaggeration of conversational structures for literary-aesthetic purposes.

Furthermore, the special grammar of oral traditions provides rich resources for the oral poet to draw from, including traditional phraseology and various thematic structures that add aesthetic interest to the storytelling. In the words of Foley, the oral poet has "a complex, heterogeneous, ever-evolving collection of inequivalent elements overseen by rules and processes no singer ever consciously imagined" to draw from in his (re-)composition. Therefore, the rich resources found in everyday conversation, including poetic features such as alliteration and prosody, can be exaggerated further in oral traditions and literature with roots in oral tradition.[50] Moreover, this exaggeration is not something that any single oral poet imagined on his/her own, but is the accumulation of generations of oral poets refining the poetic tradition as a traditional resource for future poets. In this way, the special grammar of oral poets is a type of institutional talk. Two of the examples of sequences that are analogous to ring composition given above are from institutional talk, in which multiple questions occur in one turn-at-talk followed by multiple answers in those questions in reverse order. Harvey Sacks observed the following concerning such similar phenomenon in everyday conversation: "Now it is a rather general rule that where two questions are produced, and you are going to have two answers, then the order of the answers is the reverse order of the questions".[51] The two examples above from institutional talk take this observation one step further with a series of three questions due to the

---

49 For my summary of studies of conversational storytelling, see Person, *From Conversation to Oral Tradition*, Chapter 2.

50 For my discussion of how elements in turn construction in conversation, including prosody and alliteration, provide the resources for oral poetry—for example, respectively the "heroic decasyllable" in Serbo-Croatian and the alliterative poetic line in Old English—see Person, *From Conversation to Oral Tradition*, Chapter 3.

51 Harvey Sacks, "On the Preferences for Agreement and Contiguity in Sequences in Conversation". In: Graham Button and John R. E. Lee (eds.), *Talk and Social Organization*Clevedon. 1987, 60. See Minchin, *Homeric Voices*, 110.

change in the communicative economy of the two institutions, in which question-answer sequences are the primary form of sequence—that is, Oral Proficiency Interview examinations and press conferences. Such multi-unit questioning turns "are used in institutional interactions more often than in mundane free-floating conversations".[52] Thus, if multi-turn questioning units are more common and more elaborate in such institutional settings as interviews, then it seems that multi-turn questioning units could be even more exaggerated in the institutional talk of oral poetics, especially since the (re-)performance of the tradition by any individual oral poet draws substantially from the rich resources of that tradition created by generations of oral poets. At least such an argument seems to be the case with the above example of ring composition from Homer's *Odyssey*, since it consists of a series of questions-answers in multi-unit turns, first by Odysseus and then by his dead mother (ABCDEFGGFEDCBA).

If my analysis of ring composition is accurate, then we can see how the use of ring composition in oral traditions and literature can be understood as an example of what Simpson called "foregrounding as 'more of the same'".[53] That is, those structures in everyday conversation that are analogous to ring composition can be adapted within the special grammar of oral traditions by expanding the number of parallel elements. Therefore, even if the analogous conversational structures are limited to ABCCBA patterns, the communicative economy of oral traditions allows an expansion beyond this limit, thereby producing something that attracts our aesthetic interests because of its stylistic distortion.

Hopefully, this case study of ring composition has demonstrated the utility of careful comparisons of structures identified in the special grammar of oral traditions to practices in everyday talk-in-interaction identified by conversation analysts, so that we can develop more sophisticated arguments of how a special grammar is, in Lord's words, "a grammar superimposed, as it were, on the grammar of the language concerned", so that what Jakobson identified as a "sensitive reader" and "traditional listener" "feels instinctively the poetic effect and the semantic load of these grammatical appliances".[54]

---

52 Per Linell, Johan Hofvendahl, and Camilla Lindholm, "Multi-Unit Questions in Institutional Interactions: Sequential Organizations and Communicative Functions". *Text* 23 (2003): 540. See also, Kasper ans Ross, "Multiple Questions in Oral Proficiency Interviews" and Clayman/Heritage, "Questioning Presidents".

53 Simpson, *Stylistics*, 50.

54 Lord, *Singer of Tales*, p. 36; Jakobson, *Selected Writings*, p. 93.

References

Bakker, Egbert. *Poetry in Speech: Orality and Homeric Discourse*. Ithaca 1997.

Carr, David M. *Writing on the Tablet of the Heart: Origins of Scripture and Literature*. Oxford 2005.

Claus, David B. "*Aidôs* in the Language of Achilles". *Transactions of the American Philological Association* 105 (1975): 13–28.

Clayman, Steven E. and John Heritage. "Questioning Presidents: Journalistic Deference and Adversarialness in the Press Conferences of U.S. Presidents Eisenhower and Reagan". *Journal of Communication* 52 (2002): 749–775.

Douglas, Mary. *Thinking in Circles: An Essay on Ring Composition*. Yale 2007.

Foley, John Miles. "Old English and Serbian Poetry". *Anglo-Saxon England* 12 (1983): 183–214.

Foley, John Miles (ed.). *Teaching Oral Traditions*. New York 1998.

Foley, John Miles. *Traditional Oral Epic: The Odyssey, Beowulf, and the Serbo-Croatian Return Song*. Berkeley 1990.

Jakobson, Roman. *Selected Writings: III: Poetry of Grammar and Grammar of Poetry*. The Hague 1981.

Kasper, Gabriele and Steven J. Ross. "Multiple Questions in Oral Proficiency Interviews". *Journal of Pragmatics* 39 (2007): 2045–2070.

Kendall, Calvin B. *The Metrical Grammar of Beowulf*. Cambridge 1991.

Linell, Per, Johan Hofvendahl, and Camillia Lindholm. "Multi-Unit Questions in Institutional Interactions: Sequential Organizations and Communicative Functions". *Text* 23 (2003): 539–571.

Lord, Albert. "Homer's Originality: Oral Dictated Texts". *Transactions of the American Philological Association* 84 (1953): 124–134.

Lord, Albert. "The Merging of Two Worlds: Oral and Written poetry as Carriers of Ancient Values". In: John Miles Foley (ed.), *Oral Tradition in Literature: Interpretation in Context*, 19–64. Columbia 1986.

Lord, Albert. "Ring Composition in *Mladon*; or, a Possible Case of Chiasmus in a Late Anglo-Saxon Poem". In: Joseph Harris (ed.), *The Ballad and Oral Literature*, 233–242. Cambridge 1991.

Lord, Albert. *The Singer of Tales*. Cambridge 1960.

Minchin, Elizabeth. *Homer and the Resources of Memory: Some Applications of Cognitive Theory to the* Iliad *and the* Odyssey. Oxford 2001.

Minchin, Elizabeth. *Homeric Voices: Discourse, Memory, Gender*. Oxford 2007.

Nagler, Michael N. "Towards a Generative View of the Oral Formula". *Transactions of the American Philological Association* 98 (1967): 269–311.

Niditch, Susan. *Oral World and Written Word: Ancient Israelite Literature*. Louisville 1996.

Niles, John D. *Beowulf: The Poem and Its Tradition*. Cambridge 1983.

Nimis, Steven. "The Language of Achilles: Construction vs. Representation". *Classical World* 79 (1986): 217–225.

Norrick, Neal R. *Conversational Narrative: Storytelling in Everyday Talk*. Amsterdam 2000.

Person, Raymond F., Jr. *From Conversation to Oral Tradition: A Simplest Systematics for Oral Traditions*. Studies in Stylistics and Rhetoric. London, 2016.

Person, Raymond F., Jr. *The Deuteronomic History and the Book of Chronicles: Scribal Works in an Oral World*. Atlanta 2010.

Reece, Steve. “The Three Circuits of the Suitors: A Ring Composition in *Odyssey* 17–22”. *Oral Tradition* 10 (1995): 207–229. [http://journal.oraltradition.org/issues/10i/reece]
Sacks, Harvey. “On the Preferences for Agreement and Contiguity in Sequences in Conversation”. In: Graham Button and John R. E. Lee (eds.), *Talk and Social Organization*, 54–69. Clevedon 1987.
Schegloff, Emanuel. *Sequence Organization: A Primer in Conversation Analysis 1*. Cambridge 2007.
Simpson, Paul. *Stylistics: A Resource Book for Students*. London 2004.
van Peer, Willie. “Introduction to Foregrounding”. *Language and Literature* 16 (2007): 99–106.
van Peer, Willie. “Is Literature Like a Cuckoo’s Egg?” Paper presented to the International Association for the Empirical Study of Literature. Montreal 2012.

William Duffy, University of Texas at San Antonio and
William Michael Short, University of Texas at San Antonio

Metaphor as ideology

The Greek "folk model" of the epic tradition

Abstract: Ancient Greeks' conception of the oral poetic tradition was ordered by two overarching metaphor systems that work to create an artificial opposition between epic composition and epic performance, processes that were co-extensive in actual practice. Whereas composition was understood through images of physical uprightness (especially those of "standing" or "setting up") and structural rigidity (especially those of "linking" and "binding"), performance was understood through images of flexibility and malleability (especially that of "weaving"). We propose that this metaphorically constituted "folk model" served a crucial ideological function: by distinguishing between epic-as-composition and epic-as-performance, Greek speakers were able to place their trust in the fixity and stability of the tradition as a vector of cultural knowledge, even as its performative nature entailed—and indeed depended on—it being anything but fixed and stable.

In positing that the structure of any given language—and in particular its metaphorical structure—depends in non-arbitrary ways on conceptual structure, cognitive linguistics suggests a method for reconstructing the sorts of mental models that tend to organize and guide the thinking of its speakers in most everyday contexts.[1] Metaphorical linguistic expressions reflect conventionalized mappings of more concrete (and thus more readily comprehensible) knowledge to more abstract (and thus harder to grasp) domains. Because in many cases domains are conceptualized largely (if not wholly) by means of metaphor, metaphorical mappings therefore constitute the logic and inferential patterns of those domains.[2] Further, because such mappings are by nature only partial, such metaphorically defined models can function ideologically, biasing speakers towards particular interpretations of situations or events, even when such interpretations may

1 On the formation and operation of "mental models" in human reasoning, see esp. Paul Johnson-Laird, "Mental Models in Cognitive Science". *Cognitive Science* 4 (1980): 71–115.
2 For the role of metaphorical mappings in the conveyance of logic structure, cf. Mark Johnson, *The Meaning of the Body: Aesthetics of Human Understanding*. Chicago 2008.

contradict experience.[3] In revealing the mental models that define a group's ways of interpreting experience, metaphor analysis therefore lends itself very well to the study of Greek oral poetry—where not only are the names of heroes fraught with meaning,[4] but linguistic usage in general is known to orient the audience's thinking in particular ways.[5]

In this paper, we argue that ancient Greeks' conception of the tradition of oral poetry was ordered by two overarching metaphors that work to create an artificial opposition between epic composition and epic performance. Specifically, we present evidence that the composition of epic poetry was understood through images of physical uprightness (especially those of "standing" or "setting up") and structural rigidity (especially those of "linking" and "binding"), whereas the performance of such poetry was understood through images of flexibility and malleability (especially that of "weaving"). This bifurcated metaphorical model of the tradition engenders a categorical distinction between processes that were, however, "co-extensive" in actual practice (in the sense that in the Greek oral tradition a poem's composition could not meaningfully be separated from its performance), thus establishing a framework of understanding we believe purposefully at odds with experiential reality.[6] We propose, moreover, that this "ideology" served a crucial socio-cultural function: by distinguishing epic-as-composition from epic-as-performance, Greek speakers were able to place their trust in the fixity and stability of the tradition as a vector of cultural knowledge, even as its performative nature entailed—and indeed depended on—it being anything but fixed and stable.

---

3 The papers collected in René Dirven, Rosalyn Frank, and Martin Pütz (eds.), *Cognitive Models in Language and Thought: Ideology, Metaphors and Meanings*. Berlin 2002 illustrate, in various ways, how metaphors can function ideologically; cf. also Alice Deignan, *Metaphor and Corpus Linguistics*. Amsterdam 2005, 23–24; Jacob Mey, "Metaphors and Activity". *Delta* 22 (2005): 45–65; and Terence Hawkes, *Metaphor*. London 1972.

4 Carolyn Higbie, *Heroes' Names, Heroic Identities*. New York 1995.

5 Cf. Anna Bonifazi, "Memory and Visualization in Homeric Discourse Markers". In: E. Anne Mackay (ed.), *Orality, Literacy, Memory in the Ancient Greek and Roman World*. Leiden 2007, 35–64. But see also Egbert Bakker, *Pointing at the Past: From Formula to Performance in Homeric Poetics*. Cambridge, MA 2005, who argues that certain discourse particles such as ἄρα and *δή* function to create a sense of "immediacy" by evoking visual imagery that orients the audience's understanding of what is being said.

6 Egbert Bakker, "Remembering the God's Arrival". *Arethusa* 35 (2002):. 63–81, 66; cf. John Miles Foley, "Individual Poet and Epic Tradition: Homer as Legendary Singer". *Arethusa* 31 (1998): 149–178; Gregory Nagy, "Epic as Genre". In: Margaret Beissinger, Jane Tylus, and Susanne Wofford (eds.), *Epic Traditions in the Contemporary World*. Berkeley, CA 1995, 21–32; D. Gary Miller, *Improvisation Typology, Culture and 'The New Orthodoxy'. How Oral is Homer?* Washington, DC 1982.

1 The Greek epic tradition in practice and belief

Milman Parry and Albert Lord's interviews with Serbo-Croatian *guslars* reveal a curious disconnect between the realities of oral-formulaic performance and the singer's own beliefs about this performance. Whereas observation revealed performances to be unique retellings of traditional tales using techniques designed for composing a new (if familiar) poem in performance, the *guslars* themselves maintained that they had given word-for-word reproductions of previously witnessed performances. Indeed, when the *guslars* questioned by Parry and Lord described their own compositions, they insisted, almost as a mark of professional pride, on the perfect accuracy and completeness of their songs in respect to some "original"—even if this "original" had only been heard once, and a significant interval of time previously.[7] Moreover, when confronted with the self-evident differences between other singers' and their own performances of the "same" song, they ascribed these differences to errors on the part of their rivals, rather than to any inherent feature of the performance of traditional songs.

A particularly telling example of this belief comes from an interview between Nicola Vujnović—Parry and Lord's interpreter and informant—and Đjemail (Đemo) Zogić, owner of a kafana (a kind of Serbian coffee house) in Novi Pazar where *guslars* regularly performed, as well as an accomplished singer in his own right:

> N(ICOLA). So then, last night you sang a song for us. How many times did you hear it before you were able to sing it all the way through exactly as you do now?
> Đ(EMO). Here's how many times I heard it. One Ramazan I engaged this Suljo Makić who sang for you here today those songs of the Border. I heard him one night in my coffee house. I wasn't busy. I had a waiter and he waited on my guests, and I sat down beside the singer and in one night I picked up that song. I went home, and the next night I sang it myself, and all the people said: "We would rather listen to you than to that singer whom you pay".
> N. Was it the same song, word for word, and line for line?
> Đ. The same song, word for word, and line for line. I didn't add a single line, and I didn't make a single mistake.
> N. Tell me this, if two good singers listen to a third singer who is even better, and they both boast that they can learn a song if they hear it only once, do you think that there would be any difference between the two versions?
> Đ. There would . . . It couldn't be otherwise. I told you before that two singers won't sing the same song alike.
> N. Then what are the differences?
> Đ. They add, or they make mistakes, and they forget. They don't sing every word, or they add other words.[8]

---

7 Albert Lord, *The Singer of Tales*. Cambridge, MA 2000 (first ed. 1960), 26–28.
8 Ibid., 27.

What is most notable about Zogić's statements is his simultaneous acknowledgement and dismissal of the inherent variability and adaptability of the oral poetic tradition of which he claims to be a part. While he readily acknowledges that no two singers will sing exactly the same song, he describes discrepancies between performances, most notably additions or deletions, as errors made by individual singers rather than as the result of a process of (re)composition that requires variation in the context of performance, steadfastly denying that he ever makes such mistakes in the singing of tales. Indeed, he later claims: "If I were to live for twenty years, I would sing the song which I sang for you here today just the same twenty years from now, word for word".[9]

A similar discrepancy between the realities of oral-poetic performance and discourse relating to the fidelity of transmission also characterizes the Greek epic tradition. On the one hand, we know that at an early period the poems of the epic cycle did not exist in any single fixed form. Instead, these poems were characterized by "multiformity"—that is, multiple, potentially different versions of a poem were produced over perhaps centuries of oral transmission as singers recomposed traditional tales from a stock of "formulaic" elements that were deployed variably within unique contexts of performance.[10] This multiformity was both a synchronic and a diachronic phenomenon.[11] At any given point in time, a single poem or episode in a poem could exist in many forms, each performance of it representing a distinctive instantiation of what was nevertheless recognized as the "same" song. Details of the story might also change over time: certain elements may be compressed or expanded, and even whole episodes can be omitted or reordered. Only in a later "textual" phase of Greek culture were the poems of the epic cycle sedimented in particular forms.[12]

On the other hand, poets of the Greek epic tradition consistently claim to have received—and to faithfully transmit—the tales they sing in fixed, established form. All of our extant oral Greek epics contain assertions by the poet/narrator to be simply a mouthpiece for the Muses, who are invoked as the poem's true singers, instigators of and source for the poetic act. As Ruth Scodel has shown, the Muses are seen as providing all of the information that goes into any particular performance and their influence is what sets the performance in motion, even if the "content" of the poem in question happens to have been chosen by the bard

9 Ibid.

10 Cf. Gregory Nagy, "Homeric Poetry and Problems of Multiformity: The 'Panathenaic Bottleneck'". *Classical Philology* 96 (2001): 109–119, 109–110.

11 Ibid., 110.

12 Ibid.; cf. Richard Janko, *Homer, Hesiod and the Hymns: Diachronic Development in Epic Diction*. Cambridge 1982; and Richard Janko, *The Iliad. A Commentary*. Cambridge 1992.

or audience.[13] Examples are well known. Both the *Iliad* and the *Odyssey* begin with a request—actually an imperative—that the Muse "sing" the poem: Hom. *Il.* 1.1, *μῆνιν ἄειδε θεὰ* and *Od.* 1.1, *ἄνδρα μοι ἔννεπε, μοῦσα.*[14] The poet of the *Odyssey* also commands the Muse to "utter" (1.10, *εἰπὲ*) the tale. Likewise, Hesiod's opening to the *Theogony* expresses his wish to "begin to sing from the Heliconian Muses (*μουσάων Ἑλικωνιάδων ἀρχώμεθ' ἀείδειν*)" (1) and the poet later claims that the Muses granted him both the ability and the mandate to sing about the gods:

> *ἐνέπνευσαν δέ μοι αὐδὴν*
> *θέσπιν, ἵνα κλείοιμι τά τ' ἐσσόμενα πρό τ' ἐόντα.*
> *καί μ' ἐκέλονθ' ὑμνεῖν μακάρων γένος αἰὲν ἐόντων,*
> *σφᾶς δ' αὐτὰς πρῶτόν τε καὶ ὕστατον αἰὲν ἀείδειν.* (Hes. *Th.* 31–34)
>
> And they breathed into me a divine voice,
> That I might spread fame of things about to occur and of what was before;
> And they ordered me to sing about the race of the blessed immortals,
> And to sing about them both first and last.

Here and in Homeric epic, the poet invokes the Muses not as bestowers of poetic technique and artistry, but as guarantors of the tradition's accuracy and authority. References to the Muses in epic after the initial invocation regularly center on their role in the (accurate) remembrance of information, rather than on their ability to enhance the artistry of a performance.[15] We see this at the opening of the Catalogue of Ships in *Iliad* 2, when Homer proposes to "tell the lords of the ships, and the ships' numbers" (*Il.* 2.493), but claims:

> *πληθὺν δ' οὐκ ἂν ἐγὼ μυθήσομαι οὐδ' ὀνομήνω,*
> *οὐδ' εἴ μοι δέκα μὲν γλῶσσαι, δέκα δὲ στόματ' εἶεν,*
> *φωνὴ δ' ἄρρηκτος, χάλκεον δέ μοι ἦτορ ἐνείη,*
> *εἰ μὴ Ὀλυμπιάδες Μοῦσαι Διὸς αἰγιόχοιο*
> *θυγατέρες μνησαίαθ' ὅσοι ὑπὸ Ἴλιον ἦλθον.* (Hom. *Il.* 2.488–92)
>
> I could not say or name the throng, not if there were ten tongues in me, and ten mouths, an inexhaustible tongue, and there was a brazen heart within me, if the Olympian Muses, daughters of the Aegis-bearing Zeus, did not remember who came beneath Ilion.

---

13 Ruth Scodel, "Bardic Performance and Oral Tradition in Homer". *American Journal of Philology* 119 (1998): 171–194, 173.

14 See Elizabeth Minchin, "Ring-Patterns and Ring Composition: Some Observations on the Framing of stories in Homer". *Helios* 22 (1995): 23–35.

15 Homer's *Odyssey* does not mention the Muses except in the proem (Minchin, "Ring-Patterns", 23–24) and Hesiod's *Theogony* only at the opening and closing of the work.

Note that the Muses are not credited with giving the narrator any particular physical or artistic gifts, but with remembering every single member of the fleet who came to Troy: it is the Muses, in other words, who embody the knowledge of the poet's song, and ensure that it does not go astray.[16] Emphasizing this function of recollection and transmission, four of the five other mentions of the Muses (or occurrences when the epic narrator addresses unnamed divine figures, probably identifiable as the Muses) in the *Iliad* also occur directly before lists, like that of the Trojan soldiers killed by Agamemnon in the course of his *aristeia* in Book 11.[17] The Muses, then, as guarantors of the poetic tradition's continuity and stability, appear at precisely those times when the poet breaks from direct narrative in order to provide information of a more factual nature.[18] Moreover, this is when the techniques of oral composition are most likely to introduce variation:[19] As Lord notes when discussing the presence of catalogues in Serbo-Croatian oral poetry, lists are formulaic elements especially likely to be borrowed from other, wholly different narrative traditions and adapted to suit the needs of performance.[20] Thus, the moment when an epic poet provides information that needs to appear accurate coincides with a space where the material introduced is most likely to be inconsistent or inappropriate to context. The Muses therefore provide the Greek epic tradition with incontrovertible—because divine—"proof" of its reliability as a vehicle of knowledge at the exact moment when the "text" is most lacking this quality.[21]

Oral poetry's assertion of its own trustworthiness are not limited to invocations of the Muses. Indeed, the Greek epic tradition seems to have an almost obsessive concern for the fidelity of oral-poetic transmission. When Odysseus praises the poet Demodokos at *Od.* 8.482–99, for example, he focuses not on the quality of his voice or any other aspect of his performance, but on its accuracy in respect to a sort of "original". In particular, Odysseus values that Demodokos'

16 For the occasionally shifting focus of the "Muse" within the catalogue, see Bruce Heiden, "Common People and Leaders in *Iliad* Book 2: The Invocation of the Muses and the Catalogue of Ships". *Transactions of the American Philological Association* 138 (2008): 127–154, 130.

17 Minchin, "Ring-Patterns", 27–32.

18 Cf. Benjamin Sammons, *The Rhetoric of the Homeric Catalogue*. Oxford 2010, 164–165.

19 Cf. Jim Marks, "*Arkhous Au Nêôn Ereô*: A Programmatic Function of the Iliadic Catalogue of Ships". In: Franco Montanari, Antonios Rengakos, and Christos Tsagalis (eds.), *Homeric Contexts: Neo-Analysis and the Interpretation of Oral Poetry*. Berlin 2012, 101–114.

20 Lord, *Singer of Tales*, 106. Cf. Paul Acker, *Revising Oral Theory: Formulaic Composition in Old English and Old Icelandic Verse*. New York 1997, 52–57 on the phenomenon in Old English and Icelandic epic.

21 Cf. Ruth Scodel, *Listening to Homer: Tradition, Narrative, and Audience*. Ann Arbor, MI 2009, 72–73. This is why invocations are notably absent in purely written Greek epics like Apollonius' *Argonautica*, where the textualization of the tradition provides a guarantee of its accuracy.

song has the quality of a "first-hand account (*αὐτὸς παρεὼν*)" able to capture the event "just as it happened (*κατὰ μοῖραν*)". The *Odyssey* even describes the Sirens' song more in terms of the accuracy of the knowledge it embodies than of its beauty (in contrast to virtually all other written and visual representations that focus on the beauty of their song):

> *ἴδμεν γάρ τοι πάνθ' ὅσ' ἐνὶ Τροίῃ εὐρείῃ*
> *Ἀργεῖοι Τρῶές τε θεῶν ἰότητι μόγησαν,*
> *ἴδμεν δ', ὅσσα γένηται ἐπὶ χθονὶ πουλυβοτείρῃ,* (Hom. *Od.* 12.189–91)

> We know everything which the Argives and Trojans suffered through the will of the gods in wide Troy, and we know what has happened upon the much-nourishing Earth.[22]

The invocation of the Muse(s), along with these other characterizations of poetic knowledge, function as an assertion of the oral tradition's authority *internally* to epic poetry, allowing the poet to position himself as someone who merely gives voice to the narrative. This creates a sort of "distance" between the mutable poetic performance and the supernaturally accurate epic tradition (as embodied by the Muses, Demodokos, Phemios, and so forth), which represents the song's real source and thus imbues the song with an enduring authority independent of any particular singer or any particular performance context.[23] This is the same function that the figure of Homer has *externally to* the poems, representing for Greek society a single "singer" whose authority, as (the quasi-heroic) originator and "teacher" of the oral tradition,[24] guarantees its privileged status.[25] In this sense, the relationship between poet and tradition can be described in terms of the notion of "agency" in linguistic anthropology:[26] the poet is the "animator" or immediate speaker of the poetic utterance, while the tradition is the "principal" of this utterance, representing its true origin and authority.

This is hardly an arbitrary function, given the central role that the *Iliad* and *Odyssey*—and presumably the epic tradition as a whole—played in Greek life as a

---

22 Cf. B. Pauline Nugent, "The Sounds of Sirens: *Odyssey* 12.184–91". *College Literature* 35 (2008): 45–54, 48–50.
23 Cf. Carlo Brillante, "Il cantore e la Musa nell'epica greca arcaica". *Rudiae* 4 (1993): 7–37; Claude Calame, *The Craft of Poetic Speech in Ancient Greece*. Ithaca, NY 1995.
24 Cf. Xenophanes of Kolophon, DK 21 B 10, *ἐξ ἀρχῆς καθ' Ὅμηρον, ἐπεὶ μεμαθήκασι πάντες.*
25 John Foley, *Homer's Traditional Art*. University Park, PA 2010, 56–61. In this sense, as Foley, "Individual Poet", 150 has suggested, Homer corresponds to the "legendary singer" figure in the South Slavic oral tradition, who provides "a continuing, immanent presence to the particular demands of the local context".
26 Cf. Laura Ahearn, "Language and Agency". *Annual Review of Anthropology* 30 (2001): 109–137.

source for behavioral norms, and for the conduct of social practice (even for some practices mentioned only in passing in these poems). In a very general way, for example, the Homeric texts served as a source for aristocratic social values even into the period of democracy:[27] in fact, the emergence of democracy at Athens has been traced to the persistence of a kind of "Homeric" competitiveness in aristocratic ideology.[28] Plato's *Ion* (537–41), on the other hand, refers to Homer as a specific authority on fishing, medicine, prophecy, and horsemanship.[29] Thucydides (*Hist.* 1.10.3), moreover, judged the Catalogue of Ships in the *Iliad* to be a kind of historical document, valid for argumentation about the causes of the Peloponnesian War.[30] Single lines from the Homeric epics could even be used to justify military acts or to corroborate claims of merit. Solon reportedly attempted to use the Greater Ajax's entry again in the Catalogue of Ships (Hom. *Il.* 2.557–8) in support of Athens' claim on Salamis,[31] and during the Persian War, the Athenians used Menestheus' entry (551–4) as their final and strongest piece of evidence that they should not have to yield their command to the Syracusans: Hdt. 7.161.3, *τῶν καὶ Ὅμηρος ὁ ἐποποιὸς ἄνδρα ἄριστον ἔφησε ἐς Ἴλιον ἀπικέσθαι τάξαι τε καὶ διακοσμῆσαι στρατόν*, "And the poet Homer said that [an Athenian was] the best man of those who came to Ilium in respect to drawing up and arranging an army".[32]

27 See Walter Donlan, *The Aristocratic Ideal and Selected Papers*. Wauconda, IL 1999.

28 Cf. Josiah Ober, *Mass and Elite in Democratic Athens: Rhetoric, Ideology, and the Power of the People*. Princeton, NJ 1989; see also Dean Hammer, "Ideology, Symposium, and Archaic Politics". *American Journal of Philology* 125 (2004): 479–515; and Marek Węcowski, "Homer and the Origins of the Symposium". In: Franco Montanari (ed.), *Omero Tremila Anni Dopo. Atti del congresso di Genova 6–8 luglio 2000*. Rome 2002, 625–637.

29 It is certainly possible that the *Ion*, which is primarily concerned with demonstrating the fallacy of Ion's belief that his ability to discuss Homeric epic is evidence of any true knowledge (cf. *Suzanne Stern-Gillet, "On (Mis)interpreting Plato's Ion". Phronesis 49 (2004): 169–201,* 169–172), does not mean to take seriously the notion that the *Iliad* and *Odyssey* are truly authorities on these topics. Even so, the fact that the dialogue presents this belief as existing provides evidence that classical-era Greeks believed that minute passages of the Homeric Epics provided accurate information on a wide variety of issues.

30 Thuc. *Hist.* 1.10.3, *τῇ Ὁμήρου αὖ ποιήσει εἴ τι χρὴ κἀνταῦθα πιστεύειν, ἣν εἰκὸς ἐπὶ τὸ μεῖζον μὲν ποιητὴν ὄντα κοσμῆσαι, ὅμως δὲ φαίνεται καὶ οὕτως ἐνδεεστέρα*, "If it is indeed necessary to believe the poetry of Homer, even if it was reasonable for him, being a poet, to embellish it, nevertheless [the army in the *Iliad*] appears lesser in this way".

31 Carolyn Higbie, "The Bones of a Hero, the Ashes of a Politician: Athens, Salamis, and the Usable Past". *Classical Antiquity* 16 (1997): 278–307, 283–287.

32 The attention paid to details from the Homeric epics is particularly notable in this context, given how small of a role Menestheus plays in the *Iliad* elsewhere.

The oral tradition thus served in Greek society—as oral traditions so often do cross-culturally[33]—as a repository for "cultural memory" and formed, in Eric Havelock's words, "a sort of encyclopedia of ethics, politics, history, and technology, which the effective citizen was required to learn as part of his educational equipment".[34] It also served as basis for action, conveying through time certain codes of behavior, certain exemplary stories (for imitation or moralizing), and certain "facts" about the relations between groups. For both poet and audience, the apparent stability and dependability of the tradition would therefore have been of paramount social importance. While the quality of a given performance would obviously be important to such a judgment, a successful performance would not be possible unless the poet rendered a faithful account of stories that the audience could believe to be true.[35] Furthermore, if epic poems could be used as "proof" of certain historical circumstances or relationships, or even deployed as a justification for war, those taking advantage of the tradition in this way would need to feel that this tradition was of fixed content (and not in any way modifiable or alterable on the whim of poet or audience): otherwise, any claim made on the basis of an oral poetic "text" might be invalidated simply by producing a new version of the song.[36]

2 The Greek metaphorical "folk model" of the epic tradition

If the cultural value of the Greek tradition of oral poetry was dependent on a belief in its basic stability and continuity, and at the same time multiformity was inherent to its "actualization" in (or better, as) performance, how, then, was it possible for poet and audience to perceive this tradition as anything but mutable and shifting and thus unreliable as a source of cultural knowledge?[37] As men-

33 Cf. Jan Vansina, *Oral Tradition as History*. London 1985.

34 Eric Havelock, *The Muse Learns to Write: Reflections on Orality and Literacy from Antiquity to the Present*. New Haven, CT 1963, 27.

35 Ruth Scodel, "The Story Teller and his Audience". In: Robert Fowler (ed.), *The Cambridge Companion to Homer*. Cambridge 2004, 45–55, 45.

36 In fact, Plutarch claimed that Peisistratus did exactly that with both the Hesiodic and Homeric catalog (Higbie, "Bones", 282). Since Peisistratus is credited by several ancient and modern sources as being a central figure in the standardization of epic poetry, this example is particularly apt.

37 Ruth Finnegan, *Oral Poetry: Its Nature, Significance and Social Context*. Cambridge 1977. Kahane attempts to answer this question through an argument from meter (Ahuvia Kahane,

tioned above, within epic discourse appeals to the Muse as a divine authority had a large role to play in vouchsafing the stability and continuity of the traditional knowledge this discourse was considered to contain. As Egbert Bakker has suggested, the Muses—as divinities of "remembering"—represent the natural "source" of the oral poetic tradition within "the collective mentality of a society that places fundamental authority in a remote past and considers the *κλέα ἀνδρῶν*, the record of the achievements of the ancestors that is codified in epic tradition, to be the model for excellence in the present".[38] At the same time, a cognitive linguistic perspective suggests that, at an entirely different level of representation, the metaphorical metalanguage of epic poetry also functioned to deliver a view of the tradition as "fixed" and therefore unassailable as a vehicle of custom and belief.

According to George Lakoff and Mark Johnson, understanding is frequently constituted metaphorically: that is, through the regular projection or "mapping" of more concrete, and thus more readily comprehensible concepts onto domains not directly grounded in experience.[39] Indeed, certain domains are claimed by cognitive linguists to be understood inherently by virtue of metaphor and mentally represented exclusively (or nearly so) by the metaphorical mappings that provide their structure and content. As concepts may be understood in part literally, or organized other than metaphorically (e.g., metonymically or prototypically),[40] conceptualization is thus viewed as a spectrum that runs from the more literal to the more figurative, with more concrete experiences being understood largely in their own terms and more abstract experiences understood largely in terms of others. Cognitive linguists have also shown that, particularly in the case of highly abstract domains, understanding frequently depends on multiple different figurative conceptualizations, which, though recruiting material from different sources, work together as a system to deliver an adequate overall characterization of the metaphorically defined domains. Where a concept or conceptual domain is characterized by several distinct metaphors, these metaphors normally work together, in interlocking and overlapping ways, to produce

"Hexameter Progression and the Homeric Hero's Solitary State". In: Egbert Bakker and Ahuvia Kahane (eds.), *Written Voices, Spoken Signs: Tradition, Performance, and the Epic Text*. Cambridge, MA 1997, 110–137).

38 Egbert Bakker, "Epic Remembering". In: E. Anne Mackay (ed.), *Orality, Literacy, Memory in the Ancient Greek and Roman World*. Leiden 2008, 65–78, 66.

39 George Lakoff and Mark Johnson, *Metaphors We Live By*. Chicago 1980.

40 Propositional content, image-schematic structure, conceptual metonymies, prototype effects, and so forth (Zoltán Kövecses, *Metaphors of Anger, Pride, and Love: A Lexical Approach to the Structure of Concepts*. Amsterdam 1986, 107–120).

a working conception of the domain's various aspects.[41] While the metaphors may recruit different conceptual materials (and so fail to provide a consistent overall image to conceptualization), nevertheless they typically fit together coherently as a system, each metaphor delivering an understanding of some dimension of the concept not covered, or only partially covered, by the others.[42] In this sense, metaphor systems function as the "folk models" that guide the thought and language as well as behavior in the everyday lives of language speakers, by providing a specific definition to, and so a particular basis for reasoning about, abstract domains.

Linguistic and literary evidence suggests that Greek speakers' conception of the epic tradition as a social institution was built up precisely in this way—namely, through a system of largely figurative images drawn from different domains of experience but working together to characterize this concept in its various dimensions. Consider Figure 1, illustrating the constellation of concepts converging on Greek speakers' understanding of the epic tradition. From this illustration, we see that this understanding was constituted from essentially three sources. First, the words used to refer (sometimes together, both within and without the tradition) to epic poetry most generally—namely *ἔπος* and *μῦθος*—constitute a sort of literal core of this understanding, by referring to the tradition metonymically as a product of the voice ('(POETRY >) THE TRADITION IS A VOICING'):[43] the use of both *ἔπος* (from **u̯ekʷ- "speak"*, as seen also in Greek *εἶπον*, "spoke", Latin *vox*, "voice", Sanskrit *vâcas-* and Avestan *vacah-*, "word") and *μῦθος* (likely from Egyptian *m(w)dw*, "succession of words; discourse") to define the tradition in fact reflects a common metonymic development in which the means of producing some thing comes to stand for the thing itself ('MEANS OF PRODUCING FOR PRODUCT').[44] As Richard Martin has shown, these basic understandings

---

41 On the character and function of such metaphor systems, see Marcel Danesi and Paul Perron, *Analyzing Culture*. Bloomington, IN 1999; and Zoltan Kövecses, *Language, Mind and Culture*. Oxford 2006.

42 Kövecses, *Metaphor: A Practical Introduction*. Oxford 2010, 149–167 discusses the coherence (but not consistency) of metaphor systems.

43 Cf. David Bynum, "The Generic Nature of Oral Epic Poetry". *Genre* 2 (1969): 236–258 and Ralph Rosen, "Homer and Hesiod". In: Ian Morris and Barry Powell (eds.), *New Companion to Homer*. Leiden 1997, 463–488.

44 Gunther Radden and Zoltán Kövecses, "Towards a Theory of Metonymy". In: Klaus-Uwe Panther and Gunther Radden (eds.), *Metonymy in Language and Thought*. Amsterdam 1999, 17–59. The metonymy is regular in Greek: cf., e.g., *γραφή*, "a writing; a drawing" < *γράφω*, "represent by lines" < "scratch"; *ὕφασμα*, "robe" < *ὑφαίνω*, "weave".

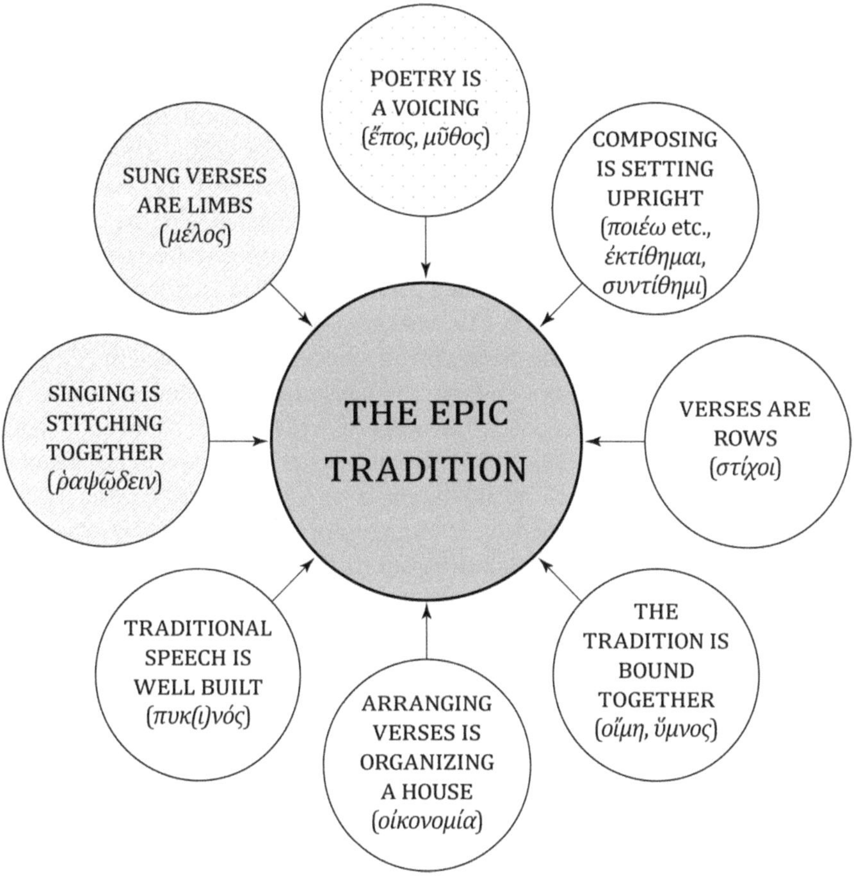

Figure 1: The Greek metaphorical "folk model" of the epic tradition.

of the tradition nevertheless bear significant (and quite different) symbolic weight: within the Homeric poems, *μῦθος*, when contrasted with *ἔπος*, designates speech, including "epic" utterance itself, that is considered authoritative or trustworthy.[45]

Beyond this skeletal literal understanding of epic poetry as a kind of "voicing",[46] however, a series of metaphorically defined concepts in Greek helps characterize the tradition in terms of rigidity, immovability, and unbrokenness. The composition of epic poetry, for example, is conventionally imagined in terms

45 Richard Martin, *The Language of Heroes: Speech and Performance in the Iliad*. Ithaca, NY 1989.
46 Walter Ong, *Orality and Literacy: The Technologizing of the Word*. London 1982, 13.

of physical uprightness: The verb *ποιέω*, commonly used to describe the creation of epic poetry, derives from Indo-European **kʷei-u-*, the basic meaning of which can be seen in the Sanskrit cognate *cinóti*, "to pile (up), erect" ... so that when Greek speakers refer to a poet of the epic tradition as, for instance, *ὁ τὴν Ἀλκμαιωνίδα πεποιηκώς*, "the author (literally, 'the builder') of the *Alcmaeonid*" or *ὁ τὴν κυκλικὴν Θηβαΐδα πεποιηκώς*, or *ὁ τὰ Κύπρια ποιήσας* or simply *ποιητής*, "the author (literally, 'the builder')", they rely on an understanding of the epic poem as a kind of upright physical construction. This image certainly seems to be implied by the terms *ἐποποιός* (cf. Suda α 3960), modelled after *ἀνδριαντοποιός*, *δρεπανοποιός*, and *μαχαιροποιός*—the "maker" (*-ποιός*) of an epic poem (*ἔπος*) being likened in this way to the makers of all kinds of concrete, material things—and *ἀοιδοθέτης*, in which the epic poet is construed as a kind of builder or craftsman who "sets up" his poem. The usage of *ἐκτίθημαι* and *συντίθημι* to refer to epic composition—as when, for example, a scholiast to Apollonius Rhodes refers to the author of the epic about Phoroneus as *ὁ δὲ Φορωνίδα συνθείς* (1.1126) —equally makes use of the image of "setting upright" (*ἐκτίθ-* < **d^heh_1-* "set up; pile up; place").[47] Dionysius of Halicarnassus (*De imit.* fr. 31.2.1) even refers to the arrangement of epic verses as *οἰκονομία*, figuring the poem as a "house", and to Homeric poetry as a statue in relief (*τῆς μὲν οὖν Ὁμηρικῆς ποιήσεως οὐ μίαν τινὰ τοῦ σώματος μοῖραν, ἀλλ' ἐκτύπωσαι τὸ σύμπαν*, "Of Homeric poetry sculpt not a single body part, as it were, but the whole ...").[48] Or verses may be *στίχες*, like any other purposefully ordered "rows" of material objects, such as trees (Xen. *Oec.* 4.21) or stone blocks (*SIG* 247 ii 72): e.g., *ἦ ῥα Μηδείας ἐπέων στίχες* (Pind. *Pyth.* 4.57).

'EPIC COMPOSITION IS SETTING UPRIGHT'

ὁ τὴν κυκλικὴν Θηβαΐδα πεποιηκώς (Athen. *Deipn.* 11.14 K)
ὁ τὰ Κύπρια Ἔπη πεποιηκὼς (15.30 K)
ὁ τὴν Θηβαΐδα πεποιηκὼς τὴν κυκλικήν (ep. 2.2, p. 54 Peppink)
ὁ τὴν Μικρὰν Ἰλιάδα πεποιηκώς (Hellan. 1a.4, fr. 152a Jacoby)
ὁ τὴν Δαναΐδα πεποιηκὼς (37.323a, fr. 27 Jacoby)
ὁ τὴν μικρὰν Ἰλιάδα πεποιηκώς (Clem. *Strom.* 1.21.104.2)
ὁ τὴν Δαναΐδα πεποιηκὼς (4.19.120.4)
ὁ τὰ Ναυπακτικὰ πεποιηκὼς (Herod. fr. 54 Müller)
Ὅμηρον τὰ ἔπη πεποιηκέναι (Iren. *Adv. haer.* 1.1.20)
τῷ τὴν Ἀλκμαιωνίδα πεποιηκότι (Dionys. fr. 3 = Schol Eur. *Orest.* 988)

47 Of material objects: *ἐκτίθημαι:* e.g., *λέβητας* (Soph. fr. 348 N); *ἆθλα* (Polyb. *Hist.* 15.9.4); etc. *συντίθημι*: e.g., *λίθους* (Thucyd. *Hist.* 4.4) *ξύλα* (Xen. *Mem.* 3.1.7); etc.

48 The word *ἐκτυπόω*, used here metaphorically of composing a Homeric epic, refers literally to modeling or working in relief: cf., e.g., Xen. *Eq.* 1.1, *ἐν τῷ βάθρῳ τὰ ἑαυτοῦ ἔργα ἐξετύπωσεν.*

τὸν πεποιηκότα τὴν κυκλικὴν Θηβαΐδα (Eustath. ad Hom. *Od.* 1 p. 413; cf. Schol. in Soph. *Oed. Col.* 1375); ὁ τὰ Κύπρια ποιήσας ἔπη (Athen. *Deipn.* 8.10 K)
ὁ τὰ Κύπρια ἔπη πεποιηκώς (15.30 K)
τὰ Κύπρια ποιήσας καὶ τὴν μικρὰν Ἰλιάδα (Corp. Aristotel. P. 1459b Bekker)
ὁ τὴν Αἰγυπτιακὴν Ἰλιάδα συνθεὶς (Athen. *Deipn.* 3.59 K)
ὁ καὶ τὰ Μεσσηνιακὰ ἔπη συνθείς (Aesch. Trag. test. 37.488 fr. 1 Jacoby)
ὁ τὰ ἔπη συνθεὶς ἃς μεγάλας Ἠοίας καλοῦσιν Ἕλληνες (Paus. 9.37.7)
ὁ τὴν Φορωνίδα συνθεὶς (Hellan. 1a.4 fr. 89 Jacoby)
καὶ τὰ ἑξῆς δὲ ἐκτίθεται σαφῶς Ὅμηρος (Eustath. ad Hom. Il. 2 p. 268)
τὸν δὲ μῦθον τοῦτον καὶ ὁ Λυκόφρων ἐκτίθεται (3 p. 133); etc.

This imagistic understanding of epic composition as a kind of physical structure may even occur within Homeric discourse, in certain metapoetic usages denoting traditional storytelling. Take the word *οἴμη* as used in Homeric texts to refer to the tales sung by a bard—as, for instance, when Demodokos is inspired to sing the tale of the Trojan War: "the Muse urged the bard to sing the glorious deeds of men, from that tale the fame of which had then reached broad heaven" (Hom. *Od.* 8.73–74, *Μοῦσ' ἄρ' ἀοιδὸν ἀνῆκεν ἀειδέμεναι κλέα ἀνδρῶν / οἴμης, τῆς τότ' ἄρα κλέος οὐρανὸν εὐρὺν ἵκανε*). Odysseus praises Demodokos in the same terms, as a possessor of traditional knowledge, "because the Muse taught you tales (*οἴμας Μοῦσ' ἐδίδαξε*)" (Hom. *Od.* 8.481). And Phemios claims that "the god inspired in my mind all kinds of tales (*οἴμας*)" (22.347). In this usage, *οἴμη* actually appears to represent the concept of the tradition of epic poetry, distinct from (but the source for) any particular instantiation of it.[49] Now, *οἴμη* has often been connected etymologically by ancient poets and modern scholars to the usage of *οἶμος*, literally "a path" or "road", to refer to "song"—poetic composition purportedly being construed in this way as a kind of journeying. For example, Pind. *Pyth.* 4.248 speaks of his narrative as a "shortcut (*οἶμον … βραχύν*)", reflecting a metaphor that is in fact widespread in many, and unrelated, oral cultures.[50] Elizabeth Minchin related this etymology to the "spatial" organization of epic poetry, but the derivation is likely spurious, determined by generic and contextual factors.[51]

49 Cf. esp. William *Thalmann*, *Conventions of Form* and Thought in Early Greek Epic Poetry. Baltimore, MD *1984*, 123–125). See also Agathe Thornton, *Homer's Iliad: Its Composition and the Motif of Supplication.* Göttingen 1984, 148–149; *Deborah Steiner, The Crown of Song: Metaphor in Pindar. London 1986*, 76–86; Andrew Ford, Homer: The Poetry of the Past. Ithaca, NY 1992, 40–48; Phoebé Giannisi, *Récits des voices. Chant et cheminement en Gréce archaïque.* Grenoble 2006.

50 Cf. Peter Agócs, *Talking Song in Early Greek Poetry.* Doctoral dissertation, University College, London 2011, 65–68.

51 Elizabeth Minchin, "Spatial Memory and the Composition of the *Iliad*". In: E. Anne Mackay (ed.), *Orality, Literacy, Memory in the Ancient Greek and Roman World.* Leiden 2008, 9–34, 28. For

Françoise Bader has argued that a more plausible etymology is the Indo-European root *seh_2-i-, "bind".[52] If this derivation is correct, *οἴμη* can then be related to Greek *ἱμάς*, denoting "leather straps" used as reins, ropes, leashes, whips, but also "planks" laid crosswise on the rafters of a roof—suggesting a metaphor of the epic tradition as a kind of physical (binding or linking) structure or again as a kind of building.

'AN EPIC SONG IS (SOMETHING) BOUND TOGETHER'

ἤδη οἴμη θελγόμενος δολιχὴν ἀνελύετ' ἄκανθαν (Apoll. Rhod. *Arg.* 4.149–50)
Μοῦσ' ἄρ' ἀοιδὸν ἀνῆκεν ἀειδέμεναι κλέα ἀνδρῶν / οἴμης, τῆς τότ' ἄρα κλέος οὐρανὸν εὐρὺν ἵκανε (Hom. *Od.* 73–74)
οἴμας Μοῦσ' ἐδίδαξε φίλησέ τε φῦλον ἀοιδῶν (8.481)
θεὸς δέ μοι ἐν φρεσὶν οἴμας / παντοίας ἐνέφυσεν (22.347)
τέρπεο δ' οἴμη / ἡμετέρη (Oppian. *Halieut.* 3.3)
δαίμοσιν εὐθήροιο μετὰ κλέος ἔρχομαι οἴμης (3.28)

Other metaphorically defined concepts of epic poetry depend on similarly "structural" or "architectural" images. Although by the fifth century *ὑμνέω* and its derivates were used in particular of lyric poetry, within epic poetry they often refer reflexively to the tradition itself—as, for example, when Homer describes Demodokos' song about the wooden horse as *ὕμνος: Od.* 8.429, *δαιτί τε τέρπηται καὶ ἀοιδῆς ὕμνον ἀκούων*, "he enjoys the meal and hearing the tale of the song". Ancient etymologies connected *ὕμνος* to *ὑπομένω*, "to remain", viewing the "hymn" as a sort durable or permanent gift to the god, "something which 'remains'" (*Etymologicum Gudianum* 540.38 Sturz),[53] or to *ὑφαίνω* (< **huph-*), viewing poetic composition metaphorically as an act of "weaving":[54] paretymological puns in fact appear already in Pindar and Bacchylides.[55] Both are problematic, however: rules of syncopation in Greek mean that the required form

Pindar's use of "road" or "path" imagery as a model for the dissemination of texts: Currie, Bruno, *Pindar and the Cult of Heroes*. Oxford 2004, 51–52. Apollo's expression comes in a poem dedicated to the god of pathways. Moreover, both occurrences of the image appear in contexts related to choral dance.

52 Françoise Bader, "Traitements de laryngales en groupe: allongement compensatoire, assimilation, anaptyxe". In: Jean Kellens (ed.), *La reconstruction des laryngales*. Paris 1990, 1–47, 36. Seen also in Old Norse *seiðr* ("kind of sorcery") and Sanskrit *sā̆man-* ("song").

53 See William Furley and Jan Bremer, *Greek Hymns*, vol. I. Tübingen 2001, 8.

54 Cf. Maria Vamvouri-Ruffy, *La Fabrique du divin. Les Hymnes de Callimaque à la lumière des Hymnes homériques et des Hymnesépigraphiques*. Kernos 2004, 19.

55 See Roland Barthes, *Le plaisir du texte*. Paris 1973, 100–101. Gregory Nagy, *Homer the Preclassic*. Berkeley, CA 2010 endorses this etymology, but admits other derivations are possible.

ὑπομονος < *ὑπομένω* cannot easily be reduced to *ὕμν-*; furthermore, the cluster **p^hn-* of the necessary preform (**huph-nos*) does not normally yield *-μν-* in Greek. A more plausible etymology might instead be **sh₂omn-os*, again from **sh₂-*, "to bind; link (together)", as proposed by Mallory and Adams.[56] Under this etymology, epic song would once again be construed as something "bound" or "linked" together, fitting very well not only with the metaphorical image of *οἴμη*, but also with the *Iliad*'s use of *πυκ(ι)νός* in reference to *ἔπος*. As Martin has argued, epic usage of *πυκ(ι)νός*—regularly used to modify such things as "arms and armor, beds, troop formations, house construction, branches, and clouds"—to describe, for example, Priam's formal request for a truce (Hom. *Il.* 7.375); Patroklos' advice to Achilles (11.788); or Zeus' command to Achilles (24.75), characterizes authoritative speech—including Homeric poetry itself—as "enduring through time, unassailable in the way of well-constructed solid, or dense-packed objects".[57]

'EPIC SONG IS WELL CONSTRUCTED' (*πυκ(ι)νός*)

ἐπέεσσιν ἀρεσσάμενος πυκινοῖσιν (Apoll. Rhod. *Arg.* 2.462)
ἡ δ' ἔπος ἐν θυμῷ πυκινὸν βάλετ' (4.1111)
Ἀρήτῃ πυκινὸν φάσθαι ἔπος Ἀλκινόοιο (4.1200)
καὶ δὲ τόδ' εἰπέμεναι πυκινὸν ἔπος (Hom. *Il.* 7.375)
ἀλλ' εὖ οἱ φάσθαι πυκινὸν ἔπος (11.788)
ὄφρά τί οἱ εἴπω πυκινὸν ἔπος (24.75)
οὐδέ τί μοι εἶπες πυκινὸν ἔπος (24.744)

To sum up: Certain discursive strategies (the Muses, "Homer" ...) are employed by Greek speakers to instill a belief in the unchangingness and contextual "independence" of the epic tradition and so to endow this tradition with a certain social and cultural authority (given that its mode of production, viz., oral performance, depends precisely on its variability and adaptability to context). Other factors also contribute to this view of the tradition by operating externally and metadiscursively—and to a large degree covertly—within Greek language and culture. In particular, we have argued, the metalanguage of epic poetry is characterized by two distinct, but related metaphorical patterns in which the composition of epic poetry is expressed in terms of "setting upright" and the tradition upon which composition draws is conveyed in terms of what is "bound together"

56 James Mallory and Douglas Adams, *Encyclopedia of Indo-European Culture.* London 1997, 520a.
57 Martin, *Language*, 35–36.

or "well built", giving an overall view of epic practice as something "rigid" and "fixed". This system of metaphors can be schematized as follows:

'EPIC COMPOSITION IS SETTING UPRIGHT'

composing an epic poem is piling (up), erecting (*ποιέω, συντίθημι, ἐκτίθημαι*)
an epic poet is a builder (*ποιητής, ἐποποιός, ἀοιδοθέτης*)
epic verses are rows (*στίχες*)
the arrangement of an epic poem is managing a house (*οἰκονομία*)
imitating an epic poem is carving a statue in relief (*ἐκτυπόω*)

'THE EPIC TRADITION IS A PHYSICAL CONSTRUCTION'

the tradition is bound together (*οἴμη*)
the tradition is solid(ly built) (*πυκινός ἔπος*)
a song is a linkage (*ὕμνος*)

We must emphasize, then, that it is not any single metaphor that produces Greek speakers' understanding of the epic tradition as "rigid" and "fixed". Indeed, since all sorts of poetic creations may be referred to in Greek by derivates of *k^wei-u-*, images of physical uprightness hardly define epic alone: heroic poems, lyric poems, dramatic poems are all equally "constructions" (*ποιῆματα*). In fact, as John Onians has pointed out, Greek speakers use "building" metaphors in their conceptualization of knowledge generally:

> [I]t was a building such as a temple which provided the Greeks with their ultimate model of secure knowledge. Already Pindar in the fifth century talks repeatedly about laying the foundation of wise words, of building a poem and of being a good craftsman or builder. At the same time *tektainomai*, 'I build,' is used for the framing of clever plans or schemes. Later when Plato seeks to establish the fixities of knowledge he develops an existing imagery and applies it with greater rigour. Knowledge, *episteme*, is literally that which you can stand on *epistamai*, just as in the *Iliad* people are said to *epistamai* on towers or walls (*Iliad*, 18, 515). To make secure knowledge it has to be built on a hypothesis, that which is 'placed' under an argument and on which it relies ... Another expression ... is *suntithemi*, 'I place together'... which is applied to thoughts and words, and which has some of the force of 'construct' ... Other examples are *mechane*, a 'contrivance' or intellectual device for achieving a goal, from *mechanaomai*, meaning originally in Homer 'I build' (*Iliad*, 8, 177).[58]

Nevertheless, the regular clustering of images of physical construction and structural integrity around concepts related specifically to notions of epic composition

58 John Onians, "Architecture, Metaphor and the Mind". *Architectural History* 35 (1992): 192–207, 200–201.

do, we believe, work together to give such a view of this part of traditional practice. In their conventional vocabulary of epic, Greek speakers cannot escape speaking of composing an epic poem as "setting up", of an epic poet as a "maker" or "builder" who "piles up" verses in "rows" in an act of "house-arrangement". At the same time, the tradition insists on a metaphorical image of itself as something that is "bound", "linked", and "well-constructed".

What has been said so far thus might appear to go against Gregory Nagy's claim that the Homeric tradition represents itself through an "esthetics of fluidity", in which the changeability of oral poetry is captured through "liquid" images.[59] Far from it. It is our claim that the "fixity" images constellating around Greek speakers' conception of epic composition in fact operate hand-in-hand with this esthetics, which focuses instead on the performative dimension of oral poetics. Indeed, the metalanguage of oral poetic performance is clearly distinguishable from that of epic composition precisely in terms of a duality that systematically opposes what is rigid and unchanging to what is flexible and changing. To sing an epic poem is *ῥαψῳδειν*, portraying this act as a "stitching together" of poetic elements—that is, as the creation of something that is textural and composite (and therefore changeable), rather than structural and integral (and therefore immutable).[60] In Plato's *Ion* (esp. 536b–c), *μέλος*, literally "member; limb", refers to the musical quality of Homeric verse as performed by a rhapsode—the term again emphasizing the "articulation" of epic song in performance.

These metaphors combine with the constellation of poetic words linked to fixity to create what Andrew Goatly calls a "hidden ideology", a (very often metaphorically defined) model that, by functioning implicitly in the language and thought of a social group, creates an understanding that may or may not coincide —and indeed may often conflict—with actual practice, or with other models guiding behavior. Of course, as Goatly points out, an ideology—meaning in the service of power—need not be "harmful", simply because it represents a "false consciousness" that contradicts material reality.[61] The ideology engendered by

---

59 Gregory Nagy, *Homer the Classic*. Hellenic Studies 36, Washington, DC 2009, 72. Nagy, "Homeric Poetry" also employs the images of "fluidity" and "rigidity" to describe the progression of poetic transmission from periods in which the tradition existed primarily orally and was characterized by decentralization, multiplicity, and variability to periods when these poems were transmitted textually, characterized by centralization, uniformity, and fixity.

60 See Derrick Collins, "Homer and Rhapsodic Competition in Performance". *Oral Tradition* 16 (2001): 129–167.

61 Andrew Goatly, *Washing the Brain: Metaphor and Hidden Ideology*. Amsterdam 2007, 1–2; cf. Teun Van Dijk, *Ideology: A Multidisciplinary Approach*. Thousand Oaks, CA 1998.

these competing metaphors, though obfuscating the true nature of oral poetic creation, serves, as we have seen, a crucial and highly adaptive social function—akin to Homeric representations of the poet as autonomous autodidacts.[62] By implying a view of epic composition as something rigid and fixed (like a building, house, or statue) and of epic performance as something flexible and fluid, where, in reality, the two processes are in Bakker's terms "co-extensive",[63] the metaphors contribute to an overall notion of the epic tradition as an absolutely authoritative and trustworthy source of cultural knowledge. Imposing, as it were, a conceptualization of the tradition that obfuscates the real process of poetic composition-in-performance, the metaphors ensure that an intrinsically mutable art form remains credible as a vehicle of cultural memory for both the poet and the audience.

3 Implications

Besides revealing the specific function that this folk model played in Greek society—namely, to characterize, in a covert way, the epic tradition as a credible vector of cultural knowledge by papering over the realities of oral poetic composition-in-performance—our analysis also has broader implications for scholars working on metaphor both in classical studies and in cognitive linguistics. To begin with, it suggests that the study of metaphor in Greek and Latin literature can move beyond the overridingly "local" view of meaning that presently defines most classicists' approach.[64] Instead of treating metaphor as a chiefly programmatic concern—that is, as related to a given author's literary program—our analysis presupposes that metaphorical structure also, or above all, pertains to a shared level of representation: in a word, to culture. Now, some scholars have recognized the role metaphor can play in constituting Greek speakers' ways of understanding the world. For instance, Jean-Pierre Vernant affirmed the existence of a "circle" metaphor that structures not only Anaximander's theory of the cosmos, but also Kleisthenes' organization of a democratic society.[65] The same image defines a certain Greek conception of the pantheon—if the image found on an altar at Epidauros (*IG* IV² 1280) is indeed meant as a sort of iconographic gloss of the inscribed dedication to the Twelve Gods.[66] Carol Dougherty has also

62 Ford, *Homer*, 90–130.

63 Bakker, "Remembering".

64 Cf. Thomas Habinek, *The Politics of Latin Literature*. Princeton 1998, 88.

65 Jean-Pierre Vernant, *The Origins of Greek Thought*. Ithaca, NY 1982.

66 That the image served this purpose is suggested by the appearance of a nearly identical symbol on altars dedicated to "all the gods and goddesses" (*IG* IV² 1, 995), "all the immortals"

studied how "marriage" afforded Greek culture a ready metaphor for conceptualizing its relationship to native peoples during colonization.[67] What is missing in these studies, however, is any attempt to link up such highly contextualized representations to understandings manifested broadly in everyday language. By demonstrating how metaphorical linguistic expressions can be used to reconstruct (equally metaphorical) models of thought, our analysis may help bridge the gap between imaginative "expert" and conventional "folk" symbolic representation, showing the basic interconnectedness of Greek speakers' signifying order.

But it is not simply a question of moving from the description of meaning at the level of language to the description of meaning at the level of culture. In classical studies, recent work along cognitive linguistic lines has already amply demonstrated that evidence from Greek and Latin can be used to reconstruct the sorts of mental models that implicitly organize ancient society's thought and behavior across various contexts of cultural practice. For our own part, William Short has demonstrated that a significant portion of Latin speakers' conceptualization of mental activity was delivered metaphorically, and in particular that "having" ideas was conceived in terms of position in physical space—with clear consequences for Roman authors' understanding of literary imitation (to give just one example of this model's effects).[68] A similar metaphorical conceptualization of MISTAKENNESS in terms of "wandering" in Latin—sharply contrasting with Greek's "missing the mark" metaphor—has been shown to motivate Roman authors' particular ways of valuing mistakes in certain pedagogical, philosophical, and artistic contexts.[69] Yet the metaphorical model reconstructed in this paper seems to have quite a different purpose. It does not provide to Greek speakers one of their automatic and unconscious cultural models that "interpret experience and guide action in a wide variety of domains including events, institutions, and physical and mental objects".[70] In fact, rather than guiding action, the model unequivocally controverts it, thus creating in those who live by

(1001, 1002), "all-gods" (1037, 1038): cf. Francesca Prescendi, "Politeismo". In: Maurizio Bettini and William Short (eds.), *Con i Romani: Un'antropologia della cultura antica*. Bologna 2014, 45–60, p. 60. See also Stella Georgoudi, "Les Douze Dieux et les autres dans l'espace culturel grec". *Kernos* 11 (1998): 73–83.

67 Carol Dougherty, The Poetics of Colonization: From City to Text in Archaic Greece. Oxford 1993.

68 William Short, "A Roman Folk Model of the Mind". *Arethusa* 45 (2012), 109–147.

69 See id., "Getting to the Truth: Metaphors of Mistakenness in Greek and Latin". *Arion* 21 (2013): 111–140.

70 Raymond Gibbs, "Taking Metaphor out of Our Heads and Putting it into the Cultural World". In: Raymond Gibbs and Gerard Steen (eds.), *Metaphor in Cognitive Linguistics*. Amsterdam 1999, 145–166, 153.

it a kind of cognitive dissonance where an operative model of understanding actively works against an operative model of practice. This dissonance, we have argued, nonetheless serves a crucial social function. In essence, the model deceives the individual for society's ends. Without wishing to reify society as an entity in itself with its own purposes different from those of its members, nevertheless this suggests that in studying metaphorically defined cultural models, scholars need to be (more) attentive to the variety of ways in which such models can operate in a social ideological sense.[71]

Furthermore, our analysis again demonstrates the fundamental importance of embodied image schemas and their metaphorical projections in human meaning construction. At the same time, it provides to cognitive linguistics much needed empirical support for the idea that conceptual metaphor operates not only synchronically but also diachronically. Perhaps justly, cognitive linguists have been criticized for their inattention to diachrony.[72] For instance, it is frequently claimed that the 'TIME IS MONEY' metaphor had its origins in the Industrial Revolution—though similar metaphors can already be found in Latin.[73] Indeed, apart from some studies describing the effects of conceptual metaphors upon a language's semantic system historically, cognitive linguists have tended to focus on the networks of metaphors that characterize concepts only synchronically.[74] By contrast, our analysis abundantly indicates the persistence of a specific set of metaphorical mappings within a social group's conceptual and linguistic system over a considerable period of time. Indeed, we have suggested that the very function (in Radcliffe-Brown's sense) of the metaphorical folk model described in this paper was to guarantee the stable transmission of cultural knowledge over time—again, by providing Greek speakers with a view of the epic tradition (and so of the cultural knowledge it contained) as rigid and fixed. If the model itself was

71 Cf. esp. Goatly, *Washing the Brain*; and James Underhill, *Creating Worldviews: Metaphor, Ideology, and Language*. Edinburgh 2011.
72 Cf. Jan Nuyts, "Cognitive Linguistics and Functional Linguistics". In: Dirk Geeraerts and Hubert Cuyckens (eds.), *The Oxford Handbook of Cognitive Linguistics*. Oxford 2007, 543–565, 550.
73 Perhaps most famously described by Lakoff and Johnson, *Metaphors*, 7–9. E.g., Vell. Pat. *Hist. Rom.* 2.89.6, *omne aevi sui spatium impensurum in id solum opus*, "about to spend the entire span of his life in that single work".
74 Notable exceptions are: Eve Sweetser, *From Etymology to Pragmatics*. Cambridge 1990; Andreas Blank, "Polysemy in the Lexicon and in Discourse". In: Brigitte Nerlich et al. (eds.), *Polysemy. Flexible Patterns of Meaning in Mind and Language*. Amsterdam 2003, 267–293; and Caroline Gevaert, "The ANGER IS HEAT Question: Detecting Cultural Influence on the Conceptualization of Anger through Diachronic Corpus Analysis". In: Nicole Delbecque, Jan van der Auwera, and Dirk Geeraerts (eds.), *Perspectives on Variation: Sociolinguistic, Historical, Comparative*. Berlin 2005, 195–208.

not also maintained and recapitulated in language diachronically, its ability to function ideologically would cease, with perhaps pernicious social effects.

Our findings may also add further nuance to the "multiple metaphor" question in conceptual metaphor theory.[75] It is very often taken as an article of faith by metaphor theorists that the different metaphorical mappings providing content and structure to a particular concept, though failing to deliver a single consistent image to conceptualization, nevertheless fit together to provide an adequate and coherent understanding of the target domain.[76] Lakoff and Johnson's explanation of the way in which the 'JOURNEY' and 'CONTAINER' metaphors converging on the concept of ARGUMENT in English cohere through shared image-schematically defined entailments is probably emblematic in this respect.[77] But less attention has been given to situations where the images metaphorically constituting a given concept conflict with one another in ways that are irreconcilable at any level of conceptual structure; still less to situations, like the one we have described, where metaphors directly contradict one another not only *within the same model* but also *as a special feature of that model*, creating a sort of bifurcated understanding in which two sets of contradictory images (RIGIDITY ↯ FLEXIBILITY) operate, pairwise, to produce contradictory conceptualizations (COMPOSITION ↯ PERFORMANCE). Cognitive linguistic studies are of course full of examples of orientational metaphors that structure conceptual relations in precisely this way, by figuring contrasting abstract concepts in terms of contrasting pairs of concrete images: e. g., 'MORE IS UP', 'LESS IS DOWN'; 'IMPORTANT IS CENTRAL', 'UNIMPORTANT IS PERIPHERAL', and so on. Ancient Greek speakers' metaphorical model of the epic tradition shows that such effects may also apply in the case of so-called "structural" metaphors characterizing more complex, culturally situated categories.[78]

Finally, our analysis indicates at least one concern for an interdisciplinary cognitive oral poetics. An important operating assumption of oral poetics has been that findings about one culture's oral tradition can be extended to those of other cultures, at least where certain formal correspondences are evident.[79] So, the understanding of how South Slavic *guslari* composed their songs "on-line"

75 See, e.g., Markus Bulgrin, *Metaphor, Or the War on Words: A Critical Analysis of the Theory of Conceptual Metaphors*. Munich 2008.

76 Kövecses, *Metaphor*, 29.

77 Lakoff and Johnson, *Metaphors*, 87–96.

78 Cf. Michael Kimmel, "Culture Regained: Situated and Compound Image Schemas". In: Beate Hampe and Joseph Grady (eds.), *From Perception to Meaning: Image Schemas in Cognitive Linguistics*. Berlin 2005, 285–311.

79 Cf., e.g., John Foley, "*Guslar* and *Aoidos*: Traditional Register in South Slavic and Homeric Epic". *Transactions of the American Philological Association* 126 (1996): 11–41.

during performance—by taking advantage of the recombinatorial possibilities of a common stock of poetic elements—could be used to explain certain features of the Homeric tradition (its reliance on formulas, frequent repetitions, inconsistencies, and so forth). In turn, insights into the nature and character of the Greek tradition could be applied to the study of, for instance, the Sanskrit *Mahābhārata*, West African *griots*, and Anglo-Saxon poems.[80] While this methodology is not unproblematic, apparently universal characteristics of oral traditions may license such generalizations.[81] Given, then, that the South Slavic oral tradition demonstrates—as we have seen—a discrepancy between practice on the one hand (performance being enabled by multiformity) and belief on the other (the professions of *guslari* vis-à-vis the fixity of composition) very much like what we have described for the Greek case, we may expect similar metaphorical patterns to occur in the metalanguage of this tradition. To our knowledge scant attention has been paid to the *guslar*'s "emic" conceptualization of composition and performance, however. Still, the brief hints given by Foley indicate thoroughgoing metaphorical structuring:[82] for instance, the verbs used by the *guslari* to denote the act of composing a song (during performance) refer literally to "enumerating" or "counting" (*brojiti*, *računati*), whereas the performance itself is understood as "driving out" or "impelling" (*turati*). The latter may be understood in relation to the physical demands of vocal performance,[83] but the contrasting images of song as the "enumeration" of a multiplex of discontinuous elements and song as a mass object requiring continuous force to "drive out" suggest a conceptualization that is equally bifurcated as—if not quite imagistically analogous with—Greek's metaphorical model. We suspect further study of the range of metaphorical ways of speaking about oral poetic traditions cross-culturally will shed light on the range of possibly different functions that such traditions may be called upon to fulfill in different societies.

80 Cf. Thomas Hale, *Griots and Griottes: Masters of Words and Music*. Bloomington, IN 2007; Francis Magoun, "Oral-Formulaic Character of A-S Narrative Poetry". *Speculum* 28 (1953): 446–467; and Robert Creed, "The Making of an Anglo-Saxon Poem". *English Literary History* 26 (1959): 445–454.

81 See Acker, *Revising Oral Theory*, 86; cf. esp. Ong, *Orality and Literacy*, 36; Albert Lord, *Epic Singers and Oral Tradition*. Ithaca, NY 1991.

82 John Foley, *Homer's Traditional Art*. University Park, PA 2010.

83 Ibid., 287, n. 40.

References

Acker, Paul. *Revising Oral Theory: Formulaic Composition in Old English and Old Icelandic Verse.* New York 1997.
Agócs, Peter. *Talking Song in Early Greek Poetry.* Doctoral dissertation, University College, London 2011.
Ahearn, Laura. "Language and Agency". *Annual Review of Anthropology* 30 (2001): 109–137.
Bader, Françoise. "Traitements de laryngales en groupe: allongement compensatoire, assimilation, anaptyxe" In: Jean Kellens (ed.), *La reconstruction des laryngales*, 1–47. Paris 1990.
Bakker, Egbert. "Remembering the God's Arrival". *Arethusa* 35 (2002): 63–81.
Bakker, Egbert. *Pointing at the Past: From Formula to Performance in Homeric Poetics.* Cambridge, MA 2005.
Bakker, Egbert. "Epic Remembering". In: E. Anne Mackay (ed.), *Orality, Literacy, Memory in the Ancient Greek and Roman World*, 65–78. Leiden 2008.
Bakker, Egbert and Ahuvia Kahane (eds.). *Written Voices, Spoken Signs: Tradition, Performance, and the Epic Text.* Cambridge, MA 1997.
Barthes, Roland, *Le plaisir du texte*. Paris 1973.
Beck, Deborah. *Homeric Conversation.* Cambridge, MA 2005.
Blank, Andreas. "Polysemy in the Lexicon and in Discourse". In: Brigitte Nerlich et al. (eds.), *Polysemy. Flexible Patterns of Meaning in Mind and Language*, 267–293. Amsterdam 2003.
Bonifazi, Anna. "Memory and Visualization in Homeric Discourse Markers". In: E. Anne Mackay (ed.), *Orality, Literacy, Memory in the Ancient Greek and Roman World*, 35–64. Leiden 2007.
Brillante, Carlo. "Il cantore e la Musa nell'epica greca arcaica". *Rudiae* 4 (1993): 7–37.
Bulgrin, Markus. *Metaphor, Or the War on Words: A Critical Analysis of the Theory of Conceptual Metaphors.* Munich 2008.
Bynum, David. "The Generic Nature of Oral Epic Poetry". *Genre* 2 (1969): 236–258.
Calame, Claude. *The Craft of Poetic Speech in Ancient Greece.* Ithaca, NY 1995.
Collins, Derrick. "Homer and Rhapsodic Competition in Performance". *Oral Tradition* 16 (2001): 129–167.
Creed, Robert. "The Making of an Anglo-Saxon Poem". *English Literary History* 26 (1959): 445–454.
Currie, Bruno. *Pindar and the Cult of Heroes.* Oxford 2004.
Danesi, Marcel and Paul Perron. *Analyzing Culture.* Bloomington, IN 1999.
Deignan, Alice. *Metaphor and Corpus Linguistics.* Amsterdam 2005.
Dirven, René, Rosalyn Frank, and Martin Pütz (eds.), *Cognitive Models in Language and Thought: Ideology, Metaphors and Meanings*. Berlin 2002.
Donlan, Walter. *The Aristocratic Ideal and Selected Papers.* Wauconda, IL 1999.
Dougherty, Carol. *The Poetics of Colonization: From City to Text in Archaic Greece.* Oxford 1993.
Dué, Casey and Mary Ebbott. *Iliad 10 and the Poetics of Ambush: A Multitext Edition with Essays and Commentary.* Washington, DC 2010.
Finnegan, Ruth. *Oral Poetry: Its Nature, Significance and Social Context.* Cambridge 1977.
Foley, John Miles. *Immanent Art: From Structure to Meaning in Traditional Oral Epic.* Bloomington, IN 1991.
Foley, John Miles. *The Singer of Tales in Performance.* Bloomington, IN 1995.
Foley, John Miles. "*Guslar* and *Aoidos*: Traditional Register in South Slavic and Homeric Epic". *Transactions of the American Philological Association* 126 (1996): 11–41.

Foley, John Miles. "Individual Poet and Epic Tradition: Homer as Legendary Singer". *Arethusa* 31 (1998): 149–178.
Foley, John Miles. *Homer's Traditional Art.* University Park, PA 2010.
Ford, Andrew. *Homer: The Poetry of the Past.* Ithaca, NY 1992.
Furley, William and Jan Bremer. *Greek Hymns.* Tübingen 2001.
Georgoudi, Stella. "Les Douze Dieux et les autres dans l'espace culturel grec". *Kernos* 11 (1998): 73–83.
Gevaert, Caroline. "The ANGER IS HEAT Question: Detecting Cultural Influence on the Conceptualization of Anger through Diachronic Corpus Analysis". In: Nicole Delbecque, Jan van der Auwera, and Dirk Geeraerts (eds.), *Perspectives on Variation: Sociolinguistic, Historical, Comparative*, 195–208. Berlin 2005.
Giannisi, Phoebé. *Récits des voices. Chant et cheminement en Gréce archaïque.* Grenoble 2006.
Gibbs, Raymond. "Taking Metaphor out of Our Heads and Putting it into the Cultural World". In: Raymond Gibbs and Gerard Steen (eds.), *Metaphor in Cognitive Linguistics*, 145–166. Amsterdam 1999.
Goatly, Andrew. *Washing the Brain: Metaphor and Hidden Ideology.* Amsterdam 2007.
Habinek, Thomas. *The Politics of Latin Literature.* Princeton 1998.
Hale, Thomas. *Griots and Griottes: Masters of Words and Music.* Bloomington, IN 2007.
Hammer, Dean. "Ideology, Symposium, and Archaic Politics". *American Journal of Philology* 125 (2004): 479–515.
Havelock, Eric. *The Muse Learns to Write: Reflections on Orality and Literacy from Antiquity to the Present.* New Haven, CT 1963.
Hawkes, Terence. *Metaphor.* London 1972.
Heiden, Bruce. "Common People and Leaders in *Iliad* Book 2: The Invocation of the Muses and the Catalogue of Ships". *Transactions of the American Philological Association* 138 (2008): 127–154.
Higbie, Carolyn. *Heroes' Names, Heroic Identities.* New York 1995.
Higbie, Carolyn. "The Bones of a Hero, the Ashes of a Politician: Athens, Salamis, and the Usable Past". *Classical Antiquity* 16 (1997): 278–307.
Janko, Richard. *Homer, Hesiod and the Hymns: Diachronic Development in Epic Diction.* Cambridge 1982.
Janko, Richard. *The Iliad. A Commentary.* Cambridge 1992.
Johnson, Mark. *The Meaning of the Body: Aesthetics of Human Understanding.* Chicago 2008.
Johnson-Laird, Paul. "Mental Models in Cognitive Science". *Cognitive Science* 4 (1980): 71–115.
Kahane, Ahuvia. "Hexameter Progression and the Homeric Hero's Solitary State". In: Egbert Bakker and Ahuvia Kahane (eds.), *Written Voices, Spoken Signs: Tradition, Performance, and the Epic Text*, 110–137. Cambridge, MA 1997.
Kimmel, Michael. "Culture Regained: Situated and Compound Image Schemas". In: Beate Hampe and Joseph Grady (eds.), *From Perception to Meaning: Image Schemas in Cognitive Linguistics*, 285–311. Berlin 2005.
Kövecses, Zoltán. *Metaphors of Anger, Pride, and Love: A Lexical Approach to the Structure of Concepts.* Amsterdam 1986.
Kövecses, Zoltán. *Language, Mind and Culture.* Oxford 2006.
Kövecses, Zoltán. *Metaphor: A Practical Introduction.* Oxford 2010.
Lakoff, George and Mark Johnson. *Metaphors We Live By.* Chicago 1980.
Lord, Albert. *The Singer of Tales.* Cambridge, MA 2000 (first ed. 1960).

Lord, Albert. *Epic Singers and Oral Tradition.* Ithaca, NY 1991.
Magoun, Francis. "Oral-Formulaic Character of A-S Narrative Poetry". *Speculum* 28 (1953): 446–467.
Mallory, James and Douglas Adams. *Encyclopedia of Indo-European Culture.* London 1997.
Marks, Jim. "*Arkhous Au Nêôn Ereô*: A Programmatic Function of the Iliadic Catalogue of Ships". In: Franco Montanari, Antonios Rengakos, and Christos Tsagalis (eds.), *Homeric Contexts: Neo-Analysis and the Interpretation of Oral Poetry*, 101–114. Berlin 2012.
Martin, Richard. *The Language of Heroes: Speech and Performance in the Iliad.* Ithaca, NY 1989.
Mey, Jacob. "Metaphors and Activity". *Delta* 22 (2005): 45–65.
Miller, D. Gary. *Improvisation Typology, Culture and 'The New Orthodoxy'. How Oral is Homer?* Washington, DC 1982.
Minchin, Elizabeth. "Ring-Patterns and Ring Composition: Some Observations on the Framing of stories in Homer". *Helios* 22 (1995): 23–35.
Minchin, Elizabeth. "The Performance of Lists and Catalogues in the Homeric Epics". In: Ian Worthington (ed.), *Voice into Text: Orality and Literacy in Ancient Greece*, 3–20. Leiden 1996.
Minchin, Elizabeth. *Homer and the Resources of Memory.* Oxford 2001.
Minchin, Elizabeth. "Spatial Memory and the Composition of the *Iliad*". In: E. Anne Mackay, (ed.), *Orality, Literacy, Memory in the Ancient Greek and Roman World*, 9–34. Leiden 2008.
Nagy, Gregory. "Epic as Genre". In: Margaret Beissinger, Jane Tylus, and Susanne Wofford (eds.), *Epic Traditions in the Contemporary World*, 21–32. Berkeley, CA 1995.
Nagy, Gregory. "Homeric Poetry and Problems of Multiformity: The 'Panathenaic Bottleneck'". *Classical Philology* 96 (2001): 109–119.
Nagy, Gregory. *Homer's Text and Language*, Champaign, IL 2004.
Nagy, Gregory. *Homer the Classic.* Hellenic Studies 36. Washington, DC 2009.
Nagy, Gregory. *Homer the Preclassic.* Berkeley, CA 2010.
Nugent, B. Pauline. "The Sounds of Sirens: *Odyssey* 12.184–91". *College Literature* 35 (2008): 45–54.
Nuyts, Jan. "Cognitive Linguistics and Functional Linguistics". In: Dirk Geeraerts and Hubert Cuyckens (eds.), *The Oxford Handbook of Cognitive Linguistics*, 543–65. Oxford 2007.
Ober, Josiah. *Mass and Elite in Democratic Athens: Rhetoric, Ideology, and the Power of the People.* Princeton, NJ 1989.
Onians, John. "Architecture, Metaphor and the Mind". *Architectural History* 35 (1992): 192–207.
Ong, Walter. *Orality and Literacy: The Technologizing of the Word.* London 1982.
Palmer, Leonard. "The Language of Homer". In: Alan Wace and Frank Stubbings (eds.), *A Companion to Homer*, 75–178. London 1962.
Prescendi, Francesca. "Politeismo". In: Maurizio Bettini and William Short (eds.), *Con i Romani: Un'antropologia della cultura antica*, 45–60. Bologna 2014.
Radden, Gunther and Zoltán Kövecses. "Towards a Theory of Metonymy". In: Klaus-Uwe Panther and Gunther Radden (eds.), *Metonymy in Language and Thought*, 17–59. Amsterdam 1999.
Richardson, Nicholas. "Homer and Oral Poetry". *Classical Review* 28 (1978): 1–2.
Rosen, Ralph. "Homer and Hesiod". In: Ian Morris and Barry Powell (eds.), *New Companion to Homer*, 463–488. Leiden 1997.
Sammons, Benjamin. *The Rhetoric of the Homeric Catalogue.* Oxford 2010.
Scodel, Ruth. "Bardic Performance and Oral Tradition in Homer". *American Journal of Philology* 119 (1998): 171–194.
Scodel, Ruth. "The Story Teller and his Audience". In: Robert Fowler (ed.), *The Cambridge Companion to Homer*, 45–55. Cambridge 2004.

Scodel, Ruth. *Listening to Homer: Tradition, Narrative, and Audience.* Ann Arbor, MI 2009.
Short, William. "A Roman Folk Model of the Mind". *Arethusa* 45 (2012): 109–147.
Short, William. "Getting to the Truth: Metaphors of Mistakenness in Greek and Latin". *Arion* 21 (2013):111–140.
Steiner, Deborah. *The Crown of Song*: Metaphor in Pindar. London 1986.
Stern-Gillet, Suzanne. "On (Mis)interpreting Plato's Ion". *Phronesis* 49 (2004): 169–201.
Sweetser, Eve. *From Etymology to Pragmatics.* Cambridge 1990.
Thalmann, William. *Conventions of Form and Thought in Early Greek Epic Poetry.* Baltimore, MD *1984.*
Thomson, John. *Studies in the Theory of Ideology.* Berkeley, CA 1984.
Thornton, Agathe. *Homer's Iliad: Its Composition and the Motif of Supplication.* Göttingen 1984.
Underhill, James. *Creating Worldviews: Metaphor, Ideology, and Language.* Edinburgh 2011.
Vamvouri-Ruffy, Maria. *La Fabrique du divin. Les Hymnes de Callimaque à la lumière des Hymnes homériques et des Hymnesépigraphiques.* Kernos 2004.
Van Dijk, Teun. *Ideology: A Multidisciplinary Approach.* Thousand Oaks, CA 1998.
Vansina, Jan. *Oral Tradition as History.* London 1985.
Vernant, Jean-Pierre. *The Origins of Greek Thought*, Ithaca, NY 1982.
Węcowski, Marek. "Homer and the Origins of the Symposium". In: Franco Montanari (ed.), *Omero Tremila Anni Dopo. Atti del congresso di Genova 6–8 luglio 2000*, 625–637. Rome 2002.

Cristóbal Pagán Cánovas, University of Navarra and
Mihailo Antović, University of Niš

Construction grammar and oral formulaic theory

Abstract: The notions of *construction* and *formula* are, respectively, the main currencies of cognitive grammar and the theory of oral-formulaic composition in performance. Even though they originated independently, both formulas and constructions are defined as form-meaning-function patterns, and, as such, represent the central theoretical constructs in their respective fields. In this chapter we propose a connection between these two research traditions, in the hope this will open up new vistas for scholars in oral poetics, cognitive linguistics, and cognitive science.

To examine how connecting these two approaches may provide new insights, we first compare the frameworks of construction grammar and the Parry-Lord theory of oral composition in performance, paying particular attention to the (often problematic) definitions of the concepts of oral formula and grammatical construction, showing that the two are based on very similar foundations. We then go on to suggest how both approaches can complement one another, outlining the connections between some of their major interests and recent developments. Finally, we illustrate our ideas with an analysis of a formula highlighted by Albert Lord in *The Singer of Tales*, using the methodological apparatus of construction grammar.

Our conclusion is that this combination of the two approaches can help us reach a better understanding of both formulas and constructions, as well as build the foundations for an interdisciplinary field of cognitive oral poetics. This new field, we hope, will have the potential to provide important insights into not only verbal art, but also into language and cognition in general, especially in the domains of generalization in language, language acquisition, and the relation between verbal form and conceptual structure.

1 The notion of formula in oral poetics: Anticipating usage-based and cognitive approaches

The notion of *formula* in oral poetics is mostly associated with the work of Milman Parry and his student Albert B. Lord (who continued their research after Parry's untimely death in 1935). The Parry-Lord theory of oral composition in performance[1] has been crucial in shaping the field of oral poetics. From the 1930s, these scholars recorded and studied a great number of epic songs in Serbo-Croatian, and compared their formulaic language with that of the *Iliad* and the *Odyssey*, proving the oral nature of the Homeric poems,[2] as well as the tradition of oral performance underlying other epics such as *The Song of Mio Cid*, *The Song of Roland*, or *Beowulf*.

Setting before themselves no smaller a task than to resolve one of the most fundamental questions in the history of literature—*whence Homer*—, Parry and Lord resorted to methods not at all typical in classical studies, in their time or even today. In his Ph.D. dissertation at the Sorbonne,[3] directed by comparative linguist Antoine Meillet (who anticipated the oral formulaic nature of Homeric poetry),[4] Parry had shown, by means of a solid statistical analysis, that Homer's *Iliad* and *Odyssey* were almost totally composed through the repetition and reuse of fixed expressions and schematic patterns for regular expressions, which he termed *formulae* and *formulaic systems*, respectively. According to Parry, this was the mark of oral composition, since the stock of ready-made chunks and *clichés* had to be a fundamental tool for an oral poet improvising during performance. Parry put forward the hypothesis that Homer was an oral poet and that the *Iliad*

1 Milman Parry, *Making of Homeric Verse: The Collected Papers of Milman Parry*. Ed. by Adam Parry. Oxford 1971; Albert Bates Lord, *The Singer of Tales*. Cambridge, MA 1960; Albert Bates Lord, *Epic Singers and Oral Tradition*. Ithaca 1991; Albert Bates Lord, *The Singer Resumes the Tale*. Ed. by Mary Louise Lord. Ithaca 1995.

2 Richard Janko, "The Homeric Poems as Oral Dictated Texts". *The Classical Quarterly* 48.1 (1998): 1–13.

3 Milman Parry, *L'épithète traditionnelle dans Homère; essai sur un problème de style homérique*. Paris 1928.

4 "L'épopée homérique est toute faite de formules que se transmettaient les poètes. Qu'on prenne un morceau quelconque, on reconnaît vite qu'il se compose de vers ou de fragments de vers qui se retrouvent textuellement dans un ou plusieurs passages. Et même les vers dont on ne retrouve pas les morceaux dans un autre passage ont aussi le caractère de formules, et ce n'est sans doute que par hasard qu'ils ne sont pas conservés ailleurs": Antoine Meillet, *Les origines indo-européennes des mètres grecs*. Paris 1923, 6.

and *Odyssey* are oral dictated texts that originated from what was once a rich, fully-oral tradition.

To gather fieldwork that would allow for a suitable comparison with the Homeric poems, Parry and his young student Albert Lord turned to the living oral epic tradition that was most easily accessible for them at the time: that of illiterate singers of pre- and post-World-War-Two Yugoslavia. The hypothesis was that, if these singers in the mounts of today's Serbia, Montenegro and Bosnia could produce pieces which at least approximated Homer in terms of length, structure and complexity, this would support the thesis that Homeric poetry was the product of an oral tradition.

In these expeditions Parry and Lord collected the materials for what would become the vast *Milman Parry Collection of Oral Literature*, hosted by Harvard University.[5] Their fieldwork and research, developed through several decades by Lord, showed that oral epic singers compose lengthy and intricate poems not by remembering a fixed text, but by improvising their song as they perform. This technique, which Parry and Lord termed "oral composition in performance", is learned through a process similar to language acquisition, and is based on the mastery of:

- *formulas*, fixed expressions regularly used under certain metrical and discursive conditions to express a given essential meaning or idea (e.g. "swift footed Achilles," "he/she spoke forth winged words," "When the child of morning, rosy-fingered Dawn, appeared", "then he mounted his horse and…")
- *themes* (also known as *type-scenes*), typical scenes or narrative blocks that structure any given story (e.g. the assembly, the arrival of a messenger, the writing of a letter, the fight between two warriors, the battle, the contest)
- *story-patterns*, a repertoire of basic plots weaving themes together (e.g. the disappearance, absence, and return of the hero, the wrath and revenge of the warrior, the abduction and rescue of the maiden)

These recurrent patterns of form, meaning, and story are much more than mere repetitions: they are the building blocks of the oral tradition. They are necessary for both memory and identity. They make the key elements of the tradition memorable, and therefore permit its preservation without written texts. They also make a performance recognizable as part of the tradition shared by poet and audience. In fact, formula density is probably the most reliable method to assess whether a given text was originated by an oral performance, whether it was

5 See http://chs119.chs.harvard.edu/mpc/.

composed by an oral traditional poet who had learned how to write, or whether it is entirely a product of literacy.[6]

Milman Parry's definition of the formula is still the most widely accepted in the field: "A group of words which is regularly employed under the same metrical conditions to express a given essential idea".[7] Thus Parry calls upon the form-meaning pairing as central to his concept of the formula, but without much specification for the meaning part. His "essential idea" has provoked many discussions among oral poeticians.

The three major aspects of the pattern are present in this definition: linguistic form, metrical slot, and meaning. However, in Parry's work the emphasis is primarily on the metrics, since he was striving to prove that the *Iliad* and the *Odyssey* were the result of oral performances. For Parry, the essential role of the formula is to provide the oral poet with a repository of set phrases that will allow him to produce appropriately-structured metrical lines during performance, while he is mentally navigating his story and anticipating what comes next.

This is no minor function. An oral poet is under great pressure during performance. In the case of the Serbo-Croatian tradition, for example, he must produce a decasyllabic line every three to six seconds.[8] Not much time to stop and think; if formulaic style is, as Parry proposed, the way in which oral epic poetry is composed and transmitted, the singer of tales will find formulas not just useful, but indispensable. Hence the great importance of formula density to assess the oral origin of a text. In Homer as well as in the Serbo-Croatian oral singers, the *guslars*, practically everything is formulaic. That is, every line, half line, or phrase between caesuras is a chunk that can also be found elsewhere, either *verbatim* or in schematic form (with its slots filled by functionally and metrically equivalent words), in the same text or group of performances by the same singer, "employed under the same metrical conditions".[9]

For this reason, at first glance at least, Parry's definition and further work on oral formulaic style could be interpreted as invoking a classic act of behavioral learning through repetition, rather than a premonition of the cognitive linguistic revolution that would come several decades later. However, as we have noted, Parry's notion of the formula, a key insight for oral poetics, included the verbal and semantic aspects as well. He privileged metrics mainly due to his immediate

6 Albert Bates Lord, "Perspectives on Recent Work on the Oral Traditional Formula". *Oral Tradition* 467 (1986): 503.

7 Milman Parry, "Studies in the Epic Technique of Oral Verse-Making. I. Homer and Homeric Style". *Harvard Studies in Classical Philology* 41 (1930): 73.

8 Lord, *The Singer of Tales*, 17.

9 For a detailed analysis of sample passages see ibid., 46–48, 142–144.

research goal. Had he not died so young (at the age of 33), there is little doubt that he would have delved into the other aspects of oral formulaic style too.

Albert Lord took up this task. His contributions expanded the semantic and functional aspects of the formula, situating it within the more complex system created by themes and story patterns. Lord's research, based on long and intensive fieldwork in South Slavic traditions as well as in the detailed comparison between Homer and medieval epic poetry, concentrates on the linguistic usage underlying this verbal art, on how the epic songs are composed, learned, and transmitted. Both empirically and theoretically, Lord anticipates some of the major tenets of cognitive grammar and usage-based approaches to language acquisition.

Lord's more complex concept of the formula[10] is already very close to constructions as defined in cognitive approaches to grammar: formulas are form-meaning pairings with a relatively fixed structure (constrained here, additionally, by metrical and prosodic rules, even some extralinguistic, such as musical factors); they are, as a rule, tolerant of some amount of formal variation (as in the Serbo-Croatian "kad eto ga," "kad eno ga," "a vidi ga," "a onda će..." which translate into English as "when here he was," "when there he was," "and then he was seen," "and then he would..."). Also, Lord shows that Parry's *formulaic systems*, abstract patterns with slots to be filled, can be extremely productive, and represent the major tool of the mature singer—much more varied and complex than mere verbatim repetition, the main resource of the apprentice. The closest concept in cognitive linguistics to these formulaic systems would likely be "schematic idioms," conventionalized phrases that can still be slightly varied. An example of such a structure could be "the X-er the Y-er," realized in idioms such as "the more, the merrier" or "the bigger the better." Here the structure seems to have a meaning of its own, e.g. "if the amount of X increases, this causes an increase in the amount of Y", yet the particular lexical choices can vary and further specify this meaning.

So metrics are indeed determinant, but they are by no means the only thing that matters.[11] For Lord, the origin of the formula is to be found in ritual repetition. The primary function of the repetitive and cumulative nature of oral formulaic style is therefore traditional: it provides a link to the time when the poet was a shaman; the metrical function, on the other hand, comes much later: "The metrical convenience, or even better, the metrical necessity, is probably a late

10 Ibid., 30–67.

11 For a more complex view of the interplay between metrical and lexical selection, see E Visser, "Formulae or Single Words? Towards a New Theory of Homeric Verse-Making", *Würzburger Jahrbücher für die Altertumswissenschaft* 14 (1988): 21–37.

phenomenon, indispensable for the growth of epic from what must have been comparatively simple narrative incantations to more complex tales intended more and more for entertainment. This was a change concomitant with the gradual shift toward the heroic and eventually the historic. It is quite likely that the later stages could not have developed until the formula became a compositional device; yet because of its past it never could become merely a compositional device. Its symbols, its sounds, its patterns were born for magic productivity, not for aesthetic satisfaction. If later they provided such satisfaction, it was only to generations which had forgotten their real meaning. The poet was sorcerer and seer before he became "artist." His structures were not abstract art, or art for its own sake."[12]

Lord's extensive experience in the field provides a detailed picture of how formulas are learned. There is no schooling or formal training in the art of epic singing. It is rather an informal apprenticeship, with coaching from the elders and innumerable hours of listening, imitation, and rehearsal in solitude. After a period as a mere listener, the young singer puts his first song together by repeatedly satisfying the various expressive needs of the story through the verbatim repetition of the formulas he can remember, until these formulas learned from others become a part of his own poetic thought. As years go by, verbatim repetition becomes less and less important, with the creative reuse of schematic phrasal patterns (of the X-killed-Y type, with abstract slots to be filled) playing the major role, and with new formulas being created through generalization from existing formulaic patterns. Thus the production of oral epic and, to a lesser extent, its reception, require the acquisition of communicative skills[13] similar to those described by usage-based approaches to language and cognition:[14] "When we speak a language, our native language, we do not repeat words and phrases we have memorized consciously, but the words and sentences emerge from habitual usage. This is true of the singer of tales working in his specialized grammar. He does not "memorize" formulas any more than we as children "memorize" language. He learns them by hearing them in other singers' songs, and by habitual usage they become part of his singing as well."[15]

---

12 Lord, *The Singer of Tales*, 67. As well as the importance of repetition in ritual, repetition plays a central role in conversation; for the parallels and differences between the functions of repetition in oral poetry and everyday speech, see Minchin (this volume).

13 Ibid., 13–29.

14 See, for example, Michael Tomasello, *Constructing a Language: A Usage-Based Theory of Language Acquisition*. Cambridge, MA 2005.

15 Lord, *The Singer of Tales*, 36.

What Lord is describing here is a usage-based acquisition process, based on repetition and resulting from a need to express a given idea with a more or less fixed form. While it is true that in this segment he was likely influenced by the epistemology of then prevalent behaviorist psychology, the emphasis on a more abstract, structural *formula*, part of the singer's semantic and poetic thought, goes far beyond classic behaviorism, giving a presentiment of cognitive linguistics. This formula is not a behavioral association, but rather a cognitively relevant structure emerging as a generalization from the process of usage. Even more importantly, this usage results in the "stabilization" of the formula, as well as in its idiolectal and dialectal variations, once more, just like in the case of natural language. Thus it seems that about 30 years before the advent of cognitive linguistics (the basis for Lord's major work, *The Singer of Tales*, is his 1948 Ph.D. dissertation), Lord's approach has many interesting connections with cognitive approaches to grammar: form-meaning templates, structural variability preserving semantic coherence, skill acquisition based on use and on richness of input, stabilization of this skill under formal constraints, online construal of the formal-semantic structures, etc.

To assess the closeness of Lordian oral formulaic theory and usage-based cognitive grammar, a cognitive linguist can do the little mental experiment of replacing *formula* and *theme* with *construction* and *frame* in the following passage from Lord's *The Singer of Tales*: "Each theme, small or large—one might even say, each formula—has around it an aura of meaning which has been put there by all the contexts in which it has occurred in the past (...) To any given poet at any given time, this meaning involves all the occasions on which he has used the theme/formula, especially those contexts in which he uses it most frequently; it involves also all the occasions on which he has heard it used by others, particularly by those singers whom he first heard in his youth, or by great singers later by whom he was impressed. To the audience the meaning of the theme involves its own experience of it as well. The communication of this suprameaning is possible because of the community of experience of poet and audience."[16]

As we can see, the idea that oral poetry, both in cognitive and communicative terms, "works like language, only more so"—as emphasized by Lord's student John Miles Foley[17]—is in many ways intrinsic to the oral formulaic theory, and was present since its very beginning. After all, Parry and Lord's fieldwork and

16 Ibid., 148.

17 John Miles Foley, *Immanent Art: From Structure to Meaning in Traditional Oral Epic*. Bloomington 1991; John Miles Foley, *How to Read an Oral Poem*. Champaign 2002, 116–145, 170–190.

comparative methodology[18] are based on the existence of cognitive universals: the human mind is expected to produce analogous solutions to attain similar communicative goals under similar conditions of performance. This is what makes the analogy between Homer and the *guslars* valid. Since oral formulaic style is acquired in a traditional, "natural" way, just like language, it must manifest, and in an enhanced fashion, the basic mechanisms employed by natural language, in all its relevant aspects: acquisition, processing, production, transmission, negotiation of meaning, interplay of patterns and creativity, etc.

More recent research on formulas has indeed incorporated a variety of aspects from linguistic theory, but, as we see, the idea of form-meaning pairings acquired as generalizations through usage remains something that researchers in oral poetics did not need to import. From the onset of the discipline, it has been clear that formulas, just like constructions in cognitive grammar and generalizations in other cognitive domains, are formed and learned on the basis of instance-based knowledge that is retained, to express it in Goldberg's terms.[19] Given these origins, it was only natural for research on oral formulaic style to develop a bigger interest in meaning and function.

Indeed, formulas are paired not only to semantic (e.g. "X killed Y", "X mounted his horse"), but also to pragmatic/functional values, which often cannot be separated from the formal or semantic dimensions (e.g. *a major character is about to speak*, *attention switches from the whole banquet to a particular character* or *the location and time of the narrative now change*). Over the last two or three decades, very relevant work has been done along these lines. Already in the eighties, we see a major change in the conception of the formula as a "ready-made phrase for anything the poet *wants to say* in favor of a ready-made phrase which accommodates what the poet wants to say to the metrical space available".[20] The major goal of this shift is to account for the creativity of the oral poet: formulas are not heavy constraints on meaning construction but "conventionalized adaptations of meaningful expressions to the metrical circumstances".[21]

18 For the Parry-Lord theory from the comparativist's point of view, see Claudio Guillén, *The Challenge of Comparative Literature*. Harvard 1993, 121, 173–178.
19 Ibid.
20 Egbert J. Bakker and Florence Fabbricotti, "Peripheral and Nuclear Semantics in Homeric Diction. The Case of Dative Expressions for 'Spear'". *Mnemosyne*, Fourth Series 44.1/2 (1991): 64.
21 Ibid., 83.

2 The notion of construction in cognitive linguistics: Connections with oral formulaic theory

Just as formulas are the main units of oral formulaic style, constructions, which are essentially formal templates paired with semantic and pragmatic content, are the building blocks of all language use, according to construction grammar. Constructions, like oral epic formulas, vary in length and structure (a morpheme, a word, a phrase, a sentence or a group of sentences), and their form, meaning and discursive function are interdependent.

Of course, the idea of analyzing syntactic structures as form-meaning pairings has a tradition of many centuries.[22] Traditional grammars from the middle ages, and in general all leading pre-Saussurean approaches to language, in the old discipline of philology, typically viewed grammatical forms as strictly related to meaning. On such views, still prevalent in elementary school education worldwide, 'nouns' *denote* objects, 'verbs' *relate to* actions, states, or events, while 'adjectives' *describe* properties. This approach was quite intuitive, as words are as a rule 'about' something in the extralinguistic reality. Thus, this 'semantic' way of thinking found its way into many more elaborate constructs of grammar, such as *exclamatory* phrases, adverbials of *manner*, or *time* clauses, to ultimately subsume entire descriptions of language. Intuitions, however, may prove to be wrong: for instance, 'earthquake' is an event, but still a noun, 'redness' a property, but not an adjective, and 'tennis' a combination of objects and events, but lacking in suffixes and therefore difficult to formally categorize as any part of speech.[23] For these reasons, and many more, the 20th century marked the period of gradual 'detachment' of grammar from semantics in most dominant linguistic schools—Saussurean and Bloomfieldean structuralism,[24] Skinnerian behaviorism,[25] and, most decisively, Chomskyan generative grammar.[26]

---

22 Hoffmann and Trousdale, *The Oxford Handbook of Construction Grammar*, vol. 2 and pages 1–14 for Hoffmann and Trousdale's introduction to the history and basic concepts of construction grammar. Valuable, though less recent summaries of the evolution of the approach are the following: Jan-Ola Östman and Mirjam Fried, *Construction Grammars: Cognitive Grounding and Theoretical Extensions.* John Benjamins Publishing, 2005), 1–15; Mirjam Fried and Jan-Ola Östman, *Construction Grammar in a Cross-Language Perspective.* Amsterdam 2004, 1–8.

23 Ray Jackendoff, *Patterns in the Mind: Language and Human Nature.* New York 1994, 68–69.

24 Ferdinand de Saussure, *Cours de linguistique générale.* Ed. by Charles Bally, Albert Sechehaye, and Tullio De Mauro. Paris 1916; Leonard Bloomfield, *Language.* Chicago 1933).

25 B. F. Skinner, *Verbal Behavior.* Acton, MA 1957.

26 Noam Chomsky, *Aspects of the Theory of Syntax.* Cambridge 1965.

So sharp was the divide between form and meaning which Chomsky proposed, that early generative linguistics ascribed practically all importance to syntactic rules, allowing semantics only a role in 'interpreting' the 'nodes' in generative trees, and even that just in relation to sentential meaning (with the entire issue of word meaning left aside). More recent schools in generative linguistics, such as Distributed Morphology,[27] Nanosyntax[28] or Lexical Meaning in Context[29] do acknowledge the importance of lexical semantics, but in a rather decompositional milieu which, even today, leaves the impression that the generative program remains faithful to its original 'syntaxocentrism'.

As with most scientific fields, paradigms in linguistics work as a pendulum. While generative approaches to syntax and semantics are still dominant in the Anglo-American study of language, in the last quarter of the twentieth century we witnessed a partial 'return' to the more traditional view of the form/meaning relationship, however in a new dress. Starting with the work of Lakoff, Langacker, and Fillmore,[30] and now fully developed within the so-called "cognitive approaches to grammar", cognitive linguistics has not merely vouched against the separation of grammar and meaning. Even more, in its attempt to revive this 'classical holism' in the science of language, but now in the modern framework of cognitive science, it has proposed the renunciation of strict, hierarchical, ultimately structuralist 'building blocks' of language, such as phonemes, morphemes, phrases, clauses or sentences. Instead of these separate labels, cognitivist schools propose that *the construction* should be (re)instituted as the central phenomenon in all language use.

So what in fact is a construction? While different authors in cognitive linguistics provide different interpretations in terms of details, some general trends are conspicuous. According to the most general definition, a construction would be a pairing of form and content. Its formal aspect is typically described as a syntactic template, but the form covers more than just syntax, as it also involves phonological aspects, as well as prosody and intonation. The content covers semantic as

---

27 Ora Matushansky and Alec Marantz (eds.), *Distributed Morphology Today*. Cambridge 2013.

28 Lambert Surhone, Mariam Tennoe, and Susan Henssonow, *Nanosyntax*. Betascript Publishing 2011.

29 Nicholas Asher. *Lexical Meaning in Context*. Cambridge 2011.

30 George Lakoff, *Women, Fire, and Dangerous Things*. Chicago 1987, 462–585; C. J. Fillmore, "Frame Semantics". In: Linguistic Society of Korea (ed.), *Linguistics in the Morning Calm*. Seoul 1982, 111–137; Charles J. Fillmore, Paul Kay, and Mary Catherine O'Connor, "Regularity and Idiomaticity in Grammatical Constructions: The Case of Let Alone". *Language* 64.3 (1988): 501–538; Ronald W. Langacker, *Foundations of Cognitive Grammar: Theoretical Prerequisites*. Stanford 1987, 1–96; For the most recent and complete overview see Hoffmann and Trousdale, *The Oxford Handbook of Construction Grammar*.

well as pragmatic meaning. In fact, any pairing of morphological, syntactic, lexical, semantic, pragmatic, or discourse-functional properties is considered a construction, while other aspects such as frequency are still subject to controversy.[31] An important immediate addendum to this rather general definition should explain that a construction must represent a *cognitively relevant structure*, rather than a mere theoretical notion: "we can say that a construction is a generalization that speakers make across a number of encounters with linguistic forms", so that a construction is always about "abstracting away" the differences in linguistic instances, seeking to construct a general pattern."[32]

One can obviously satisfy the criteria from the definition above in a number of ways (depending ultimately on one's exact view of linguistic 'form' and 'content', which can of course tremendously differ). This indeed allows for a number of approaches to constructions by individual authors, which has perhaps prompted Turner and Steen to remark that "construction grammarians use the term 'construction' in a range of ways, but in all of them, a construction is a mental packet, consisting of a form-meaning pairing that speakers of a language know."[33] While most authors agree on this much, individual positions usually branch into two directions—very broad and very specific ones, both of which we mention in further text.

As a good example of the broad view of constructions in linguistics, one may suggest Langacker's Cognitive Grammar, an influential school calling for a recategorization of practically our entire knowledge of traditional or formal grammar, in line with general principles of perceptual and cognitive psychology. Langacker defines constructions as follows: "Most of the expressions we employ are symbolically complex, to some degree analyzable into smaller symbolic elements. Grammar consists of the patterns for constructing such expressions. Accordingly, the expressions and the patterns are referred to as constructions."[34] In perhaps the most liberal definition of all that we have found, the same author adds that "a construction is defined as either an expression (of any size) or else a schema abstracted from expressions to capture their commonality (at any level of specificity)."[35] Similarly, Diessel defines constructions as "the basic units of grammar (...) commonly defined as grammatical assemblies that are characterized by two features: first, constructions combine a specific form with a specific meaning

31 For a panorama of current views, see Hoffmann and Trousdale, *The Oxford Handbook of Construction Grammar*, 1–14.

32 Martin Hilpert, *Construction Grammar and Its Application to English*. Edinburgh 2014.

33 Steen and Turner, "Multimodal Construction Grammar," 4.

34 Ronald W. Langacker, *Cognitive Grammar: A Basic Introduction*. Oxford 2008, 161.

35 Ronald W. Langacker, *Investigations in Cognitive Grammar*. Berlin and New York 2009, 2.

(...); and second, constructions exhibit both general grammatical properties and idiosyncratic features."[36] For Geeraerts,[37] "a grammatical construction is any string of words or morphemes exhibiting a coherent pattern, whether it be an entire sentence or a clause or a phrase (like a noun phrase or a verb phrase) or a complex lexeme (like a phrasal verb)."

Thus, definitions of this first kind stress the connection between form and meaning, the need for this connection to be cognitively relevant, and the resulting 'obsolescence' of more traditional classifications into grammatical categories, such as morphemes, clauses, or sentences. How broad the definition is seems to depend on the scope of language phenomena that an author wishes to encompass.[38]

While retaining these broad methodological assumptions, a number of scholars support somewhat narrower views of constructions, appropriate to their particular fields of interest. For Croft and Cruse constructions satisfy all criteria defined above, but are still restricted to 'grammatical units' *larger* than individual words (which excludes the traditional domains of phonology and many segments of morphology).[39] Thus their 'radical construction grammar' focuses on what more structuralist approaches to language would call syntax. Also, in a related development, construction grammar and much of cognitive linguistics originated from the study of idioms—expressions which seem to have a complex syntactic form, but indecomposable semantic content, and are thus difficult to account for in formal models, such as generative grammar. As Goldberg claims "unlike mainstream generative grammar, constructionist approaches tend to emphasize the detailed semantics and distribution of particular words, grammatical morphemes, and cross-linguistically unusual phrasal patterns; the hypothesis behind this methodology is that an account of the rich semantic/pragmatic and complex formal constraints on these patterns readily extends to more general, simple, regular patterns."[40]

In our opinion, the "middle ground" between these two contrasting groups of definitions—quite general and rather particular—can be found exactly in the work

---

36 Holger Diessel, *The Acquisition of Complex Sentences*. Cambridge 2004, 14.

37 Dirk Geeraerts, "A Rough Guide to Cognitive Linguistics". In: Dirk Geeraerts (ed), *Cognitive Linguistics: Basic Readings*. Berlin and New York 2006, 16..

38 For instance, one may quote here Croft who seems to implicitly complain that "It is a short step from analyzing the resultative construction [as realized in sentences such as *He wiped the plate clean*] as a construction to analyzing all the syntactic rules of a language as constructions." (Croft 2007, 468).

39 William Croft and D. Allan Cruise, *Cognitive Linguistics*. Cambridge 2004, 291–328.

40 Goldberg, *Constructions at Work*, 5.

of Adele Goldberg: "(...) basic sentences of English are instances of constructions —form-meaning correspondences that exist independently of particular verbs. (...) particular semantic structures together with their associated formal expression must be recognized as constructions independent of the lexical items which instantiate them."[41] Or, more algorithmically put a few pages later: "C is a construction iff C is a form-meaning pair (F, S), such that some aspect of F, or some aspect of S is not strictly predictable from C's component parts or from other previously established constructions. That is, a construction can be posited in the grammar if it can be shown that its meaning and/or its form is not compositionally derived from other constructions existing in the language."[42] And finally: "A construction is (...) a pairing of form with meaning/use such that some aspect of the form or some aspect of the meaning use is not strictly predictable from the component parts or from other constructions already established to exist in the language".[43] Besides emergent meaning, frequency is the other parameter that can make a form-meaning pair qualify as a construction.[44]

One may resort to a few often-invoked examples to better illustrate the kind of form-meaning connection that cognitive linguists have in mind when looking for constructions. A typical such example is the "caused motion construction", as given in the English sentence *He blew the napkin off the table.* The argument is that the meaning of the expression does not emerge only from the senses and references of individual words used, especially "content words" (blow, napkin, table), but that the *structure itself* also carries a portion of this meaning—where any X causing any Y off something denotes that X is exerting a force onto Y so that the Y changes its location. Since the structure contains part of the semantics, the construction retains its meaning even if the sense of the main verb would not typically imply any causation, as in "*They laughed the poor guy out of the room.*"[45] Likewise, the already-mentioned construction "the -er, the -er" carries some semantics regardless of the particular adjectives used—such as "the more, the merrier". Thus, any traditional grammatical form, at any level, which provides a gist of possible meaning even before it is "filled" with particular lexical items—may qualify as a construction.

Therefore, with all above definitions in mind, we can attempt to summarize the minimum conditions a construction needs to satisfy. In order for a linguistic

41 Adele E. Goldberg, *Constructions: A Construction Grammar Approach to Argument Structure.* Chicago 1995, 1.

42 Ibid., 4.

43 Ibid., 68.

44 Goldberg, *Constructions at Work.*

45 Cf. Goldberg, *Constructions*, 152.

structure to potentially qualify as a construction it should be (1) a cognitively relevant phenomenon; (2) one which cannot be obviously reduced to simpler structures; and (3) one in which form and meaning are inextricably related—in the sense that the form itself carries some "schematic" meaning, while the words with their senses, "rich in conceptual content",[46] serve to further specify the final meaning of the expression. Importantly, (4) the "form" does not need to refer only to syntax, but may be related to phenomena traditionally ascribed to morphology and phonology, including meter, or even suprasegmental elements, such as linguistic and paralinguistic prosodic features (changes in pitch, timbre, etc.).

Thus, we hope to have shown that, as theoretical constructs, the formula and the construction share some relevant constituent elements, even if the particular terms used by the two research communities are somewhat different. In the next section, we provide a sample analysis to show how the notions of formula and construction may be brought together.

3 Analysis: The horse-mounting pattern in Serbo-Croatian epic

Now we provide an analysis whose goal is to show that formulaic segments may be interpreted as constructions in the cognitive-linguistic sense of the word.[47] The example we will use is one that Lord himself found very interesting. In *The Singer of Tales* (p. 52) he quotes the formula which we provisionally label "the horse-mounting pattern". This sequence repeats a number of times in different songs, and also in performances sung by different singers. It represents a typical situation in which a hero mounts his horse in order to set out for a journey or, even more proverbially, a battle. Here are some of the variations that Lord notes:

- Pa zasede krilata đogata. [Then he mounted his winged horse.]
- Pa posede šajku bedeviju. [Then he mounted his bedouin mare.]
- Eh, zasede njezina đogata. [Well, she mounted her white horse.]
- Zasedoše konje u avliju. [They mounted their horses in the courtyard.]
- Pa zaseše konje na jaliju. [Then they mounted their horses on the bank.]
- Zasedoše dva konja menzila. [They mounted two post horses.]

46 Vyvyan Evans and Melanie Green, *Cognitive linguistics: An Introduction*. Mahwah, NJ 2006, 525.

47 For further aspects of construction grammar and frame semantics that can be applied to the analysis of poetic texts, see Boas (this volume).

In the given chapter of the book, Lord provides an extensive discussion on the realizations of this formula, insisting that no individual singer can know all the variations, that it is impossible to determine who was the first to devise one version of the formula or the other, and that there is no "handbook" of formulae to select from: they rather belong to a "common stock" (p. 49). He goes on to explain the importance of heroes' horses in the Yugoslav tradition and then lists a number of appearances of the particular formula, typically in the songs sung by Salih Ugljanin, a bilingual Albanian and Serbo-Croatian singer from the town of Novi Pazar, present-day Serbia, with whom he was working at the time.

However, while he does admit that the nature of the formula is somewhat 'abstract', Lord does not go much beyond the metrical analysis of its structure. Typically, he provides syllabic and metrical justifications for the use of a particular verb or combination of verbs, prepositions and nouns, and discusses implications for the makeup of the verse. For instance (p. 52, note the use of the term "idea" to include both grammatical and semantic aspects of lexical items): "When a modifier is added to the idea 'horse', or when an adverbial idea is to be added to the idea 'mounted', the verb moves to the first half line. Or, to state it in another way, if the verb is put in the first half line, some modifier must be added to the idea 'horse' or an adverbial idea must be added to the idea 'mounted'". While this analysis is certainly important for Lord's purpose—to uncover potential metrical constraints underlying formulaic variation—he still seems to miss a key point in these examples—the connection between the use of the grammatical structure and the schematic meaning of the utterances. Thus, as stated in the previous section, while he masterfully provided the overall epistemological background precursing cognitive linguistics in the early chapters of his book, Lord unfortunately neglected meaning in most of his individual analyses of formulae. Accordingly, it seems to us that there is a lot of room left for a cognitive linguistic analysis that uses the notion of a construction.

As can be seen in the lines quoted, in all examples we find variations of the verb "sedeti" ("sit"). The verb is used in the third person singular or plural, with the subject omitted, which is possible in Serbo-Croatian, since verb endings are inflected. Although its principal meaning is indeed "sit," Lord, quite rightly, and probably intuitively, translates the entire structure as containing the verb "mount." In his analysis directed at verse and metrics, he pays no attention to this apparent suppletion, yet it is central to the analysis here. The meaning of the verb is indeed "to mount," yet not because of the verb's dictionary meaning, its sense ("Sinn"), but rather because of the grammatical structure in which the verb is used, which itself carries a portion of the verb's "meaning."

The translation of the Serbo-Croatian "sedeti" ["sit"] into English as "mount" is possible for three reasons: first, in all six cases, there is *a prefix* before the main verb (za-sede, po-sede, za-sedoše, po-sedoše, za-seše). Both "za-" and "po-" used as verb prefixes typically denote a *commencement* of an action (as in "zaspati"—fall asleep, as opposed to "spavati"—(be) sleep(ing); or in "poželeti"—make a wish, as opposed to "želeti"—wish, want, reflecting the *process* of wishing). Therefore, the use of the prefixes implies that the sitting is not a state of any durability, but a momentary action. Second, in terms of the tense used, what we have here is the *aorist*, which appears in all six cases, in the third-person singular twice and then third-person plural three times. There has been quite a lot of discussion on the particular role, use and grammatical meaning of the aorist in Serbo-Croatian.[48] Most authors seem to agree that the aorist implies the completion of a short, typically recent, action (some also add its pragmatic effects, such as the unexpected nature of the action, to which the speaker also has an emotional attitude; Arsenijević also provides significant evidence that the aorist should be labeled a "null tense"). So, the "grammatical meaning" of the particular combination of the prefixes and the aorist may be translated as "He/They *started sitting*." The sitting commences, takes a very short while, it is recent, and, in some sense, emotional for the narrator. In other words, in synergy, both the aorist and the prefix denote the "beginning" of an action, and provide a strong perfective aspect. In Serbo-Croat, the perfectivity is obviously resolved on the level of what would traditionally be called morphology, while in English, the same resolution would need to be syntactic or semantic. In other words, "On je sedeo" means "He was sitting." "On zaseše" would be translated into English as "He sat," or, even more precisely, though of course periphrastically, "He commenced a short action which results in the new state—that of sitting." This structural discrepancy between the two languages is what most likely motivated the apparent suppletion in the English translation, in which the hero "mounts" the horse, rather than "sits" on one.

And this is not all. When the metrical structure allows it—in the particular example, when the verb is disyllabic (third person singular aorist—*zasede,* third person plural aorist—*zaseše*), rather than trisyllabic (third person plural aorist doublet—*zasedoše*)—there is an empty "slot" remaining, which needs to be filled with a monosyllabic word so that the singer could retain the decasyllabic meter. The typical choice that Ugljanin makes is "pa." Parry, and perhaps also Lord, would have probably claimed that this short word ("idea") has no meaning and

---

48 For excellent recent positions, which also offer useful reviews, see , Mateusz-Milan Stanojević and Renata Geld, "Epistemic immediacy and the aorist". In: Adeline Patard and Frank Brisard (eds.), *Cognitive Approaches to Tense, Aspect, and Epistemic Modality*. Amsterdam 2011, 159–180 (for Croatian); and Boban Arsenijević, "The tense and aspect components in the semantics of the /Serbian/ aorist". *Srpski jezik* 18 (2013): 253–261 (for Serbian).

that it makes its way into the verse only for metrical reasons. However, our position is that this lexical item should be viewed not as a meaningless particle, but rather as a full-fledged conjunction with temporal meaning (as in the complex sentence: "Uradio je domaći *pa* je izašao napolje"—"He did his homework *and then* he went outside."). Thus "pa" suggests that the hero had been doing something else, "and then" moved on to the new action, that of mounting a horse.

In other words, if we take into account *all three grammatical phenomena*—the use of the aorist, the verb prefixes, and the two out of six instances of the temporal "pa" before the main verb (when the meter allows this), *we may claim that there is a schematic meaning in the structure itself*, which not only turns the sense of the verb "sit" into that of the verb "mount", but also provides a *more general idea of introducing a beginning of a new action, which suddenly commences, replacing a previous one*. Among other things, this approach can go beyond Lord's metrical analysis in at least one important respect: it can provide a stronger explanation of the *motivation* for the singer's six-fold use of the particular formula, amidst the plentitude of other logically, grammatically and metrically possible combinations, which could also nicely fit the general message, or "idea", that a hero has mounted his horse.

The proposed structure thus represents a form-meaning pair, which one may algorithmically define as follows:

> FORM: ([optional temporal conjunction / filler] + verb with prefixation + aorist + noun/ object + [optional attribute to the object / spatial adverbial]
>
> MEANING: (a perfective, momentary action which has just started, immediately after the previous action finished, to which the speaker has an emotional attitude) + the object of the action + optional description of the object

Thus, it looks to us that it is *the structure itself which has a schematic meaning*. The aorist verb has a prefix denoting the beginning of the action and perfectivity. In addition, when the metrics allows it, this is intensified by a filler/temporal adverbial. When the lexical items are added, this meaning is then additionally "filled", or made "richer in conceptual content", by invoking the notion of *mounting* and *horses*, with all of their lexical senses and references, relevant to the narration that follows.

In other words, when there is no prefix or the aorist, sitting on a horse is a *process* (like in English continuous tenses); with the prefix and the aorist, sitting on the horse is the *actual momentary action of jumping onto one* (like in simple or simple perfect tenses). Even more handily, if another syllable is necessary for metrical reasons, this prefixed verb is additionally preceded by a filler- temporal deictic/conjunction ("pa!"), which serves to intensify the semantics of the prefix, where the compositional meaning is, roughly, "and then he changed his state by mounting a horse."

While the particularities of the metrics themselves provide interesting insights, we hope to have shown in this short example that one may apply the analysis further than Lord, where this kind of structure strongly resembles the cognitive linguistic notion of construction.

4 Conclusions: The usage-based approach to language and verbal art

We have proposed that construction grammar and oral formulaic theory are very congenial frameworks. Seeking to explain language in general and oral poetics in particular, each of them has produced very similar analytic categories and tools. It is precisely their close attention to actual use that has led them through their respective paths. Crucially, both approaches focus on the idiomatic aspects of linguistic performance, showing how these seemingly peripheral elements are in fact central: idiomaticity is the way language works, from the leisure of everyday conversation to the enormous demands of traditional epic performance. And this is probably due to universals of cognition, to commonalities in how the human mind generalizes over instances, abstracts useful patterns, and reuses those templates for ad-hoc purposes.

Constructionist approaches can provide a theoretical framework to study oral formulaic style as a window into the mind, linking it to general meaning construction and cognitive processes. Such a framework would be based on the following tenets, which are shared by most construction grammars:[49] (1) language is articulated through learned pairings of form and function with a (mainly) phrasal nature, (2) grammar does not involve transformation or derivation and semantics is associated directly with surface form, (3) linguistic knowledge includes both items and generalizations over instances, acquired through a usage-based process. Therefore, utterances are usually novel and formulaic at the same time, since constructions place great constraints on lexical choice and at the same time have open slots that can be filled with novel lexical content.

49 Taken from Adele Goldberg, "Constructionist Approaches," in *The Oxford Handbook of Construction Grammar*, ed. Thomas Hoffmann and Graeme Trousdale (Oxford; New York: Oxford University Press, 2013), 15–31.

Oral formulaic style exposes all these aspects and provides a great living laboratory for their study, due to the extreme idiomaticity necessitated by composition in performance. At the same time, oral formulaic theory calls the attention of the linguist to the role of tradition and performance in the construction of linguistic meaning, suggesting that construction grammars could pay more attention to diachronic and communicative factors.

A new discipline of cognitive oral poetics would need to integrate the methodologies and insights of both oral poetics and cognitive science, and particularly cognitive linguistics. We hope to have shown in this paper that two central concepts of oral composition in performance and cognitive grammar, the formula and the construction, are based on very similar epistemologies. Likewise, we also hope to have shown how the analysis of formulas as constructions, with the use of the tools from cognitive linguistics, may prove to be deeper and more appropriate than what Parry and Lord proposed more than half a century ago, in that they provide the necessary link between the form and meaning of the poetic performance, but also an insight into the cognitive processes in the mind of the singer as he sings. It is our intuition that this will be only one of many connections in the emerging field of cognitive oral poetics.

References

Arsenijević, Boban. “The tense and aspect components in the semantics of the /Serbian/ aorist”. *Srpski jezik* 18 (2013): 253–261.

Asher, Nicholas. *Lexical Meaning in Context*. Cambridge 2011.

Bakker, Egbert J. and Florence Fabbricotti. “Peripheral and Nuclear Semantics in Homeric Diction. The Case of Dative Expressions for ‘Spear’”. *Mnemosyne*, Fourth Series, 44.1/2 (1991): 63–84.

Bloomfield, Leonard. *Language*. Chicago 1933.

Chomsky, Noam. *Aspects of the Theory of Syntax*. Cambridge 1965.

Evans, Vyvyan and Melanie Green. *Cognitive linguistics: An Introduction*. Mahwah, NJ 2006.

Fillmore, Charles J., Paul Kay, and Mary Catherine O’Connor. “Regularity and Idiomaticity in Grammatical Constructions: The Case of Let Alone”. *Language* 64.3 (1988): 501–538.

Fillmore, C. J. “Frame Semantics”. In: Linguistic Society of Korea (ed.), *Linguistics in the Morning Calm*, 111–137. Seoul 1982.

Foley, John Miles. *How to Read an Oral Poem*. Champaign 2002.

Foley, John Miles. *Immanent Art: From Structure to Meaning in Traditional Oral Epic*. Bloomington 1991.

Fried, Mirjam and Jan-Ola Östman. *Construction Grammar in a Cross-Language Perspective*. Amsterdam 2004.

Goldberg, Adele. “Constructionist Approaches”. In: Thomas Hoffmann and Graeme Trousdale (eds.), *The Oxford Handbook of Construction Grammar*, 15–31. New York 2013.

Goldberg, Adele. “The Nature of Generalization in Language”. *Cognitive Linguistics* 20.1 (2009): 93–127.
Goldberg, Adele. *Constructions at Work: The Nature of Generalization in Language*. Oxford 2006.
Goldberg, Adele. *Constructions: A Construction Grammar Approach to Argument Structure*. Chicago 1995.
Guillén, Claudio. *The Challenge of Comparative Literature*. Cambridge 1993.
Hoffmann, Thomas and Graeme Trousdale (eds.). *The Oxford Handbook of Construction Grammar*. Oxford and New York 2013.
Jackendoff, Ray. *Patterns in the Mind: Language and Human Nature*. New York 1994.
Janko, Richard. “The Homeric Poems as Oral Dictated Texts”. *The Classical Quarterly* 48.1 (1998): 1–13.
Lakoff, George. *Women, Fire, and Dangerous Things*. Chicago 1987.
Langacker, Ronald W. *Investigations in Cognitive Grammar*. Berlin and New York 2009.
Langacker, Ronald W. *Cognitive Grammar: A Basic Introduction*. New York 2008.
Langacker, Ronald W. *Foundations of Cognitive Grammar: Theoretical Prerequisites*. Stanford 1987.
Lord, Albert B. *The Singer Resumes the Tale*. Ed. by Mary Louise Lord. 1 edn. Ithaca 1995.
Lord, Albert B. *Epic Singers and Oral Tradition*. Ithaka 1991.
Lord, Albert B. “Perspectives on Recent Work on the Oral Traditional Formula”. *Oral Tradition* 467 (1986): 503.
Lord, Albert B. *The Singer of Tales*. Cambridge, MA 1960.
Meillet, Antoine. *Les origines indo-européennes des mètres grecs*. Paris 1923.
Östman, Jan-Ola and Mirjam Fried. *Construction Grammars: Cognitive Grounding and Theoretical Extensions*. Amsterdam 2005.
Parry, Milman. *Making of Homeric Verse: The Collected Papers of Milman Parry*. Ed. by Adam Parry. Oxford 1971.
Parry, Milman. “Studies in the Epic Technique of Oral Verse-Making. I. Homer and Homeric Style”. *Harvard Studies in Classical Philology* 41 (1930): 73.
Parry, Milman. *L’épithète traditionnelle dans Homère; essai sur un problème de style homérique*. Paris 1928.
Saussure, Ferdinand de. *Cours de linguistique générale*. Ed. by Charles Bally, Albert Sechehaye, and Tullio De Mauro. Paris 1916.
Skinner, B. F. *Verbal Behavior*. Acton, MA 1957.
Stanojević, Mateusz-Milan and Renata Geld. “Epistemic immediacy and the aorist”. In: Adeline Patard and Frank Brisard (eds.), *Cognitive Approaches to Tense, Aspect, and Epistemic Modality*, 159–180. Amsterdam 2011.
Steen, Francis and Mark B. Turner. “Multimodal Construction Grammar”. In: Michael Borkent, Barbara Dancygier, and Jennifer Hinnell (eds.), *Language and the Creative Mind*. Stanford, CA 2013.
Surhone, Lambert, Mariam Tennoe, and Susan Henssonow. *Nanosyntax*. Betascript Publishing, 2011.
Tomasello, Michael. *Constructing a Language: A Usage-Based Theory of Language Acquisition*. Cambridge, MA 2005.
Visser, E. “Formulae or Single Words? Towards a New Theory of Homeric Verse-Making”. *Würzburger Jahrbücher für die Altertumswissenschaft* 14 (1988): 21–37.

Hans C. Boas (Austin)

Frames and constructions for the study of oral poetics

Abstract: The seminal studies by Milman Parry and Albert Lord in the 20th century demonstrated that oral epic singers compose lengthy and intricate poems not by remembering a fixed text, but by improvising their songs as they perform. Such performances rely heavily on mastering a set of interconnected concepts, such as (1) formulae, (2) themes, and (3) story-patterns, all of which need to be accessed throughout a performance. In communicating with their audience, oral poets rely on features of everyday language use, which is often dependent on pairings of form with meaning and encyclopedic knowledge. The formulas used by oral poets constitute a very high degree of idiomaticity, which is often difficult to analyze systematically with the concepts of more traditional grammatical theories.

In this paper, I show how two related linguistic theories, Construction Grammar and Frame Semantics, can be applied to the study of oral poetics in order to systematically describe and analyze the forms and meanings communicated by oral poets during their performances. The paper is structured as follows. Section 1 discusses the principles of Frame Semantics, a theory of word meanings. It then shows how FrameNet, a lexicographic database of English, is structured to represent different types of lexical knowledge important for the interpretation of words. Section 2 presents Construction Grammar, a framework for systematically analyzing syntactic structures together with their meanings. Section 3 shows how the principles of Frame Semantics and Construction Grammar are applied for the analysis of a concrete text. Section 4 summarizes the main points of this paper.

1 Frame Semantics and FrameNet

One central problem with analyzing language is figuring out the meanings of words. Traditional print dictionaries often offer good first characterizations of word meanings, but they suffer from a number of drawbacks, such as circularity, mode of organization, level of detail, and coverage.[1] Over the past five decades,

1 See Beryl T.S. Atkins, "Analyzing the Verbs of Seeing: A Frame Semantic Approach to Corpus Lexicography". In: Christopher Johnson et al. (eds.), *Proceedings of the Twentieth Annual Meeting of the Berkeley Linguistics Society*, Berkeley 1994, 42–56; Beryl T.S. Atkins, "The Role of the Example in a Frame Semantics Dictionary". In: Masayoshi Shibatani and Sandra A. Thompson

linguists developed different types of lexical semantic theories to account for word meanings, including structuralist semantics, generativist semantics, neo-structuralist semantics, and cognitive semantics.[2]

One very prominent cognitive semantics approach, developed by Charles Fillmore in the 1970s and 1980s, is Frame Semantics, "a research program in empirical semantics and a descriptive framework for presenting the results of such research."[3] This approach relies on common backgrounds of knowledge (semantic frames) against which the meanings of words are interpreted.[4] A "frame is a cognitive structuring device, parts of which are indexed by words associated with it and used in the service of understanding."[5] The central ideas underlying Frame Semantics can be characterized as follows:

> A word's meaning can be understood only with reference to a structured background of experience, beliefs, or practices, constituting a kind of conceptual prerequisite for understanding the meaning. Speakers can be said to know the meaning of the word only by first understanding the background frames that motivate the concept that the word encodes. Within such an approach, words or word senses are not related to each other directly, word to word, but only by way of their links to common background frames and indications of the manner in which their meanings highlight particular elements of such frames.[6]

(eds.), *Essays in Semantics and Pragmatics—In Honor of Charles J. Fillmore*. Amsterdam and Philadelphia 1995, 25–42; Beryl T.S. Atkins and Michael Rundell, *The Oxford Guide to Lexicography*. Oxford 2008; and Cliff Goddard, *Semantic Analysis*. Oxford 2011.

2 For an overview, see Vyvyan Evans, *How words mean*. Oxford 2009; and Dirk Geeraerts, *Theories of Lexical Semantics*. Oxford 2010.

3 Charles J. Fillmore, "Frame Semantics". In: Linguistic Society of Korea (ed.), *Linguistics in the Morning Calm*. Seoul 1982, 111–138, 111. This section builds on Hans C. Boas, "Recent trends in multilingual computational lexicography". In: H. C. Boas (ed.), *Multilingual FrameNets in Computational Lexicography*. Berlin and New York 2009, 1–36; Hans C. Boas, "Zum Abstraktionsgrad von Resultativkonstruktionen". In: S. Engelberg, K. Proost, and A. Holler (eds.), *Sprachliches Wissen zwischen Lexikon und Grammatik*. Berlin and New York 2011, 37–69; and Hans C. Boas, "Wie viel Wissen steckt in Wörterbüchern? Eine frame-semantische Perspektive". In: *Zeitschrift für Angewandte Linguistik* 57 (2013): 75–97.

4 Fillmore's use of the concept of "frame" is related to other research in linguistics and artificial intelligences. For details, see Alexander Ziem, *Frames und sprachliches Wissen*. Berlin and New York 2008; and Dietrich Busse, *Frame Semantik*. Berlin and New York 2012.

5 Miriam R.L. Petruck, "Frame Semantics". in: J. Blommaert, C. Bulcaen, J.-O. Östman, and J. Verschueren (eds.), *Handbook of Pragmatics*. Amsterdam 1996, 1–13, 2.

6 Charles J. Fillmore and Beryl T.S. Atkins, "Toward a Frame-based Lexicon: The Semantics of RISK and its Neighbors". In: E. Kittay and A. Lehrer (eds.), *Frames, Fields and Contrasts: New Essays in Semantic and Lexical Organization*. Hillsdale 1992, 75–102, 76–77. For a more detailed review of the main principles of Frame Semantics, see Petruck, "Frame Semantics"; Charles J. Fillmore, Chris R. Johnson, and Miriam R.L. Petruck, "Background to FrameNet". *International Journal of Lexicography* 16 (2003): 235–251; and Charles J. Fillmore and Collin Baker, "A frames

To illustrate, consider the so-called Revenge frame, which is evoked by a number of semantically related words such as the verbs *revenge*, *avenge*, *get even*, the nouns *revenger*, *sanction*, and *retaliation*, and the adjectives *vengeful*, and *vindictive*.[7] The Revenge frame represents a scenario with different participants, also known as frame elements (FEs) that can be regarded as instances of broader semantic roles such as Agent, Undergoer, Instrument, etc.[8] FEs are defined specifically with respect to the types of situations covered by frames.[9] In FrameNet (http://framenet.icsi.berkeley.edu), an on-line lexicographic database implementing the principle of Frame Semantics, the Revenge frame is defined as follows:

> The frame concerns the infliction of punishment in return for a wrong suffered. An Avenger performs a Punishment on an Offender as a consequence of an earlier action by the Offender, the Injury. The Avenger inflicting the Punishment need not be the same as the Injured_Party who suffered the Injury, but the Avenger does have to share the judgment that the Offender's action was wrong. The judgment that the Offender had inflicted an Injury is made without regard to the law.

I first discuss the so-called core FEs of the Revenge frame as defined by FrameNet. The Avenger is defined as someone who exacts revenge from the Offender for the Injury. The FE Injured_Party identifies the constituent that encodes who or what suffered the Injury at the hands of the Offender. Sometimes, an abstract concept such as a person's honor or their blood is represented as the element that has suffered the injury. These also constitute instances of Injured_Party. The FE Injury is the injurious action committed by the Offender

---

approach to semantic analysis". in: B. Heine and H. Narrog (eds.), *The Oxford Handbook of Linguistic Analysis*. Oxford 2010, 313–340.

7 Names of semantic frames are in Courier font. Names of Frame Elements (FEs) are in small caps. Frame Elements differ from traditional universal semantic (or thematic) roles such as Agent or Patient in that they are specific to the frame in which they are used to describe participants in certain types of scenarios. "Tgt" stands for target word, which is the word that evokes the semantic frame.

8 See Robert D. Van Valin, Jr. and D. P. Wilkins, "The Case for 'Effector': Case Roles, Agents, and Agency Revisited". In: M. Shibatani/S. Thompson (eds.), *Grammatical Constructions*, Oxford 1996, 289–322; Fillmore et al., "Background"; Hans C. Boas, "Semantic frames as interlingual representations for multilingual lexical databases". *International Journal of Lexicography* 18.4 (2005): 445–478; Fillmore and Baker, "A frames approach"; and Josef Ruppenhofer, Michael Ellsworth, Miriam R. L. Petruck, Chris Johnson, and Jan Scheffczyk, *FrameNet II: Extended theory and practice*. 2010. Available at [http://framenet.icsi.berkeley.edu].

9 See also Thomas Schmidt, "The Kicktionary—A multilingual lexical resource of football language". In: H. C. Boas (ed.), *Multilingual FrameNets: Methods and Applications*. Berlin and New York 2009, 101–134, on the difference between frames and scenarios.

against the Injured_Party. This FE need not always be realized, although it is conceptually necessary. The Offender has committed the earlier Injury for which the Avenger seeks revenge. The FE Punishment is defined as the action taken by the Avenger, i.e. the Avenger carries out a Punishment in order to exact revenge on the Offender.

Besides the core FEs Avenger, Injured_Party, Injury, Offender, and Punishment, there are also a number of non-core FEs belonging to the `Revenge` frame. Non-core FEs are more peripheral FEs that are not always conceptually necessary for the scenario described by a frame. They typically include FEs such as Degree (which identifies the Degree to which an event occurs), Instrument (i.e. the Instrument with which the revenge is performed), Manner (which refers to the Manner in which the Avenger exacts their revenge from the Offender), Place (which identifies the Place where the revenge occurs), Purpose (this FE identifies the Purpose for which the revenge is performed), and Time (which identifies the Time when the revenge occurs).

With this background in mind, I now turn to some concrete examples of how the `Revenge` frame is evoked. Frame Semantics subscribes to a splitting approach (as opposed to a lumping approach) when analyzing the various senses of a word. On this view, the number of a word's senses depends on the number of semantic frames it evokes. Each frame-evoking sense of a word is called a lexical unit (LU), and each lexical unit evokes a separate semantic frame. In the following examples, the frame-evoking LUs are in all caps to indicate their special status.

(1) [$_{\text{Avenger}}$ Bubba] AVENGED^{Tgt} [$_{\text{Injury}}$ the death of his cat] [$_{\text{Punishment}}$ by killing] [$_{\text{Offender}}$ the coyote].

(2) [$_{\text{Avenger}}$ I] shall take a terrible RETRIBUTION^{Tgt} upon [$_{\text{Offender}}$ you] [$_{\text{Injury}}$ for all the pain you have caused me].

(3) They look upon the security forces as a VENGEFUL^{Tgt} [$_{\text{Avenger}}$ occupying force which must be opposed].

In (1)-(3), the `Revenge` frame is evoked by three different target LUs (identified the superscript "Tgt"), namely the verbal LU *avenge*, the nominal LU *retribution*, and the adjectival LU *vengeful*. The annotations in the three sentences are the result of applying the description of the `Revenge` frame to the sentences. For example, in (1) the verbal LU *avenge* evokes the `Revenge` frame, and the core FEs, which are realized syntactically, include the Avenger (*Bubba*), the Injury (*the death of his cat*), the Punishment (*by killing*), and the Offender (*the coyote*).

The `Revenge` frame is just one of thousands of frames discovered, documented, and analyzed by frame semanticists over the past four decades. One major advantage of semantic frames is that they allow us to systematically catalogue the types of knowledge evoked by words in specific contexts. There are many different types of frames at different levels of abstraction,[10] from many different semantic domains, covering various types of concepts such as Activity_Stop, Age, Calendric_Unit, Theft, Operate_Vehicle, Part_Whole, Quarreling, Departing, Suspicion, Temperature, and Waiting. Since each frame has its own unique set of FEs it becomes possible to systematically compare and contrast word meanings. It also allows us to arrive at inferences about particular scenarios even in the absence of lexically specific information. Consider, for example, the sentence *We saw the presents under the tree*, which clearly evokes the Christmas frame, together with a particular specification with respect to the Calendric_Unit frame (though this is heavily culture dependent, Christmas presents in the U.S. are typically opened on the morning of December 25, while in Germany this traditionally happens on December 24), and associated frames such as Religious_practices and Holidays.

Another advantage is that frames can be systematically related to each other by so-called frame-to-frame relations.[11] This allows us to connect related types of information to each other. For example, the LU *to mug* evokes the Robbery frame, which uses a more abstract frame, the Theft frame, and it inherits from the frame Committing_crime. Connecting one frame to other frames allows us to systematically understand how concepts are related to each other, and how very often they depend on each other. See, for example, Fillmore and Atkins[12] on the Risk frame, Petruck and Boas[13] on the Calendric_Unit frame in English, German, and Hebrew, and Schmidt[14] on numerous frames important for the analysis of word meanings in the soccer domain.

One major advantage not addressed so far concerns the relationship of form to meaning. As I show below, Frame Semantics is not only interested in accounting for the meanings of words. It is also concerned with how that meaning is syntactically realized and under what circumstances specific elements of meaning (i.e. FEs) can be left unsaid. I return to this point below.

The examples in (1)-(3) illustrate how some of the basic principles of Frame Semantics are applied to the analysis of word meanings. But how is such knowledge helpful for our understanding of Oral Poetics? Recall that during their performances, oral poets rely heavily on mastering a set of interconnected concepts, such as (1) formulae, (2) themes, and (3) story-patterns. But so far, there

---

10 See Fillmore and Baker, "A frames approach".

11 See Ruppenhofer et al., *FrameNet II*.

12 Fillmore and Atkins, "Frame-based Lexicon".

13 Miriam R.L. Petruck and Hans C. Boas, "All in a Day's Week". In: E. Hajicova, A. Kotesovcova, and Jiri Mirovsky (eds.), *Proceedings of CIL 17*. Prague 2003.

14 Schmidt, "Kicktionary".

does not yet exist a systematic approach for cataloguing and analyzing the types of knowledge accessed by oral poets during their performances. I propose that Frame Semantics offers an important solution that will help us with systematically describing and analyzing the different types of knowledge accessed by oral poets. To this end, I now turn to a more detailed discussion of how the principles of Frame Semantics have been applied to the creation of FrameNet, a database of lexical entries for several thousand English words taken from a variety of semantic domains.

Based on corpus data from the British National Corpus, members of the FrameNet project identify and describe semantic frames and analyze the meanings of words by appealing directly to the frames that underlie their meanings.[15] In addition, the FrameNet project studies the syntactic properties of words by asking how their semantic properties are given syntactic form.[16] Between 1997 and 2015, FrameNet defined close to 13,000 lexical units (LUs) (a word in one of its senses) in more than 1,100 frames.

The workflow of FrameNet begins by defining frame descriptions (based on corpus evidence) for the words to be analyzed. Then, the following steps are taken:

> (1) characterizing schematically the kind of entity or situation represented by the frame, (2) choosing mnemonics for labeling the entities or components of the frame, and (3) constructing a working list of words that appear to belong to the frame, where membership in the same frame will mean that the phrases that contain the LUs will all permit comparable semantic analyses.[17]

The next step focuses on finding corpus sentences in the British National Corpus that illustrate typical uses of the target words in specific frames. These corpus sentences are then extracted mechanically and annotated manually by tagging the FEs realized in them. Finally, lexical entries are automatically prepared and stored in the database.[18]

---

15 FrameNet is usage-based, i.e. it deals with naturally occurring (British) English from a more general domain of English (90% written (different genres) and 10% spoken language). There exist specific FrameNet analyses for the medical and judicial domains, where language use is quite different. Taking the usage-based approach to the domain of oral poetics would require linguists to do systematic corpus studies of different texts to see what types of semantic frames are evoked by the words (lexical units) in them. In the case of old texts this might be rather complicated because we usually do not have any native speaker intuitions about all the frame-specific knowledge from the time when the texts were originally written.

16 Fillmore et al., "Background", 235.

17 Charles J. Fillmore, Miriam R.L. Petruck, Josef Ruppenhofer, and Abby Wright, "FrameNet in Action: The Case of Attaching". *International Journal of Lexicography* 16 (2003): 297–333, 297.

18 For more details, see Fillmore and Baker, "A frames approach".

The result of this workflow is an online dictionary of English that is structured in terms of semantic frames. Going to the FrameNet website, users can search—among other things—for entries of specific LUs, frame descriptions, and combinations thereof. Lexical entries in FrameNet offer a link to the definition of the frame evoked by a LU, including FE definitions, and example sentences exemplifying prototypical instances of FEs. In addition, FrameNet includes a list of all LUs evoking the same frame while also providing frame-specific information about various frame-to-frame relations, like the child-parent relation and sub-frame relation.[19] For example, a search for the `Revenge` frame returns a frame description, together with a list of several semantically related words such as *retaliate, avenge, avenger,* and *revengeful*, among others, all of which evoke the same frame (for the frame definition (including the FEs), see above).

Consider the LU *retaliate*, whose FrameNet entry consists of three parts. FrameNet uses different colors to highlight each FE, making it easier to identify individual FEs. Due to formatting restrictions, FE names are not color-coded in this paper as they are in the on-line FrameNet entries.

Figure 1 illustrates the first part of a lexical entry in FrameNet, namely the Realization Table of the Lexical Entry Report. Besides providing a dictionary definition of the relevant LU, in this case *retaliate*, it summarizes the different syntactic realizations of the FEs. The left column lists the names of different core FEs (AVENGER, INJURED_PARTY, INJURY, and INSTRUMENT), the middle column lists the number of annotated example sentences in FrameNet, and the right column lists the different types of syntactic realizations of the respective FEs. Consider the FE INJURY, which appears 38 times, 35 of those times as a definite null instantiation (DNI),[20] twice as a dependent prepositional phrase headed by *against*, and once by a dependent propositional phrase headed by *for*.

The second part of the Lexical Entry Report summarizes the valence patterns found with a LU, that is, "the various combinations of frame elements and their syntactic realizations which might be present in a given sentence."[21] This list presents an exhaustive summary of all frame element configurations found in the corpus, i.e. combinations of FEs occurring together in the same sentence, together with their various syntactic realizations. While some frame element configurations

---

19 Fillmore et al., "FrameNet in Action".

20 Following Charles J. Fillmore, "Pragmatically controlled zero anaphora". In: *Proceedings of the Berkeley Linguistics Society*, 95–107. Berkeley 1986, FrameNet records which FEs can be null-instantiated, that is which FEs do not have to be overtly realized syntactically. There are three different types of null instantiation: indefinite null instantiation (INI), definite null instantiation (DNI), and constructional null instantiation (CNI).

21 Fillmore et al., "Background", 330.

Frame Element	Number Annotated	Realization(s)
Avenger	(39)	CNI.-- (2) NP.Ext (37)
Injured_party	(1)	PP[on].Dep (1)
Injury	(38)	DNI.-- (35) PP[against].Dep (2) PP[for].Dep (1)
Instrument	(3)	PP[with].Dep (3)
Manner	(1)	AVP.Dep (1)
Offender	(39)	DNI.-- (36) PP[against].Dep (2) PP[on].Dep (1)
Place	(1)	PP[at].Dep (1)
Punishment	(39)	PP[in].Dep (2) PP[with].Dep (3) INI.-- (19) AVP.Dep (2) PPing[by].Dep (12) DNI.-- (1)
Time	(2)	AVP.Dep (2)

Figure 1: First part of FrameNet entry for *retaliate*

display only one specific syntactic realization pattern, such as that in the first line AVENGER, INJURED_PARTY, OFFENDER, and PUNISHMENT), others exhibit a greater syntactic variability, such as that of AVENGER, INJURY, OFFENDER, and PUNISHMENT, which has a total of twelve different syntactic realizations. The second column from the left in the valence table for *retaliate* in Figure 2 illustrates how the FE AVENGER may be realized either as an external noun phrase (NP.Ext) or not at all, in the case of constructional null instantiation (CNI). Clicking on the link in the column to the left of the valence patterns leads the user to a display of annotated examples sentences illustrating the valence pattern, as exemplified in (1)-(3)

Number Annotated	Patterns				
1 TOTAL	Avenger	Injured party	Offender	Punishment	
(1)	NP Ext	PP[on] Dep	DNI --	INI --	
3 TOTAL	Avenger	Injury	Instrument	Offender	Punishment
(3)	NP Ext	DNI --	PP[with] Dep	DNI --	INI --
1 TOTAL	Avenger	Injury	Manner	Offender	Punishment
(1)	NP Ext	DNI --	AVP Dep	DNI --	INI --
1 TOTAL	Avenger	Injury	Offender	Place	Punishment
(1)	NP Ext	DNI --	DNI --	PP[at] Dep	PP[with] Dep
31 TOTAL	Avenger	Injury	Offender	Punishment	
(1)	CNI --	DNI --	DNI --	PP[in] Dep	
(1)	CNI --	DNI --	DNI --	PP[with] Dep	
(2)	NP Ext	DNI --	DNI --	AVP Dep	
(9)	NP Ext	DNI --	DNI --	INI --	
(1)	NP Ext	DNI --	DNI --	PP[in] Dep	
(1)	NP Ext	DNI --	DNI --	PP[with] Dep	
(10)	NP Ext	DNI --	DNI --	PPing[by] Dep	
(2)	NP Ext	DNI --	PP[against] Dep	INI --	
(1)	NP Ext	DNI --	PP[on] Dep	DNI --	
(1)	NP Ext	PP[against] Dep	DNI --	INI --	
(1)	NP Ext	PP[against] Dep	DNI --	PPing[by] Dep	
(1)	NP Ext	PP[for] Dep	DNI --	INI --	
2 TOTAL	Avenger	Injury	Offender	Punishment	Time
(1)	NP Ext	DNI --	DNI --	INI --	AVP Dep
(1)	NP Ext	DNI --	DNI --	PPing[by] Dep	AVP Dep

Figure 2: Second part of FrameNet entry for *retaliate*, Valence Table (partial excerpt)

above. It is important to remember that the list of annotated example sentences is the result of the manual semantic annotation process completed by FrameNet researchers before the actual lexical entry can be compiled. As such, users can always consult the full list of semantically annotated example sentences that form the basis of the Realization Table and the Valence Table in Figures 1 and 2 above.[22]

Having summarized the main principles of Frame Semantics, we are now in a better position to answer the question of how it can help us with understanding the mechanisms and concepts of oral poetry. Recall that all oral poets have more or less the same themes and story-patterns in their minds: the hero that returns home, the hero that dies in the place of another hero, the deity that replaces a hero in disguise, etc. These themes are essentially small scripts or building blocks that help the poet build a part of one of these patterns, e.g. *the messenger, the assembly, the battle, the fight between two warriors, the pursuit*, etc. Themes can be re-used in all or most story-patterns. Looking at themes from a frame-semantic point of view, one could say that themes are organized sets of traditional knowledge shard by the oral poet and the audience. On this view, specific types of scripts (a.k.a. frames) are evoked by specific themes (a.k.a. frame-evoking LUs (which can also be multi-word expressions)).

For example, if a particular story pattern involves a script for a battle, then the text needs to include a particular frame-evoking LU such as *altercation, battle, clash, combat*, or *fight*, which evokes the `Hostile_encounter` frame. Once this frame is evoked, the oral poet (and the audience) know that there have to be a number of specific FEs involved, such as an Issue (an unresolved question over which the two sides of a hostile encounter are in disagreement), a Purpose (the desired result of the outcome of the hostile encounter for the Side_1 or for all the Sides), a Side_1 (one of two participants in a hostile encounter), a Side_2 (one of two participants in a hostile encounter), and various other non-core FEs such as Degree, Duration, Manner, Means, Place, Time, etc. Whether or not all FEs are indeed realized syntactically in sentences describing a battle is of course lexically determined by the properties of individual LUs. But whether they are realized syntactically or not in a given context does not matter, since both the oral poet and its audience presumably share the same frame-semantic knowledge about what is involved in battles. In other words, modeling themes with semantic frames also allows us to account for variation in a poet's performance.

22 For more information, see Hans C. Boas, "From theory to practice: Frame Semantics and the design of FrameNet". In: S. Langer and D. Schnorbusch (eds.), *Semantik im Lexikon*. Tübingen 2005, 129–160; and Boas, "Recent trends".

On some occasion the poet might chose to evoke the battle theme (a.k.a. the `Hostile_encounter` frame) with the noun *battle*. On other occasions, the poet might instead choose *clash*, *combat*, *confrontation*, or *fight*. Even though these are different words with slightly different meanings, they all evoke the same semantic frame, and using different paraphrases may allow the poet to achieve different stylistic effects depending on the occasion. This strategy also cuts across different parts of speech. Instead of using the nouns *battle* and *fight*, the poet might choose the verbal counterparts *to battle* and *to fight*, or support verb constructions such as *to pick a fight* or *to enter battle*.

As explained above, the frame-semantic approach can also capture much more complex relationships in texts. Consider, e.g., the various frames connected to `Hostile_Encounter` that are also at least indirectly evoked when a LU evokes the `Hostile_Encounter` frame. These include frames such as `Fighting_activity` (*affray*, *fray*, *melee*), `Attack` (*assault*, *ambush*, *charge*, *fire*, *onslaught*, etc.), `Taking_sides` (*against*, *endorse*, *oppose*, *opposition*, etc.), `Friendly_or_hostile` (*enemy*, *friendly*, *hostile*), and `Member_of_military` (*soldier*, *troop*, *general*). This means that once the poet evokes the `Hostile_encounter` frame by using a particular LU such as *fight* or *battle*, this not only triggers all the relevant knowledge associated with such events, but it may also lead to evoking related frames which can serve to elaborate on the theme or the overall script.[23] Using the frame-based approach to investigate oral poetry thus allows us to systematically catalogue the types of traditional knowledge shared by the performer and the audience.

2 Construction Grammar

So far, I have mainly addressed how meaning is relevant for the study of oral poetry. I now turn to the question of how the various meanings are expressed syntactically. This involves several factors. The first factor, which I already discussed in the context of valence patterns above, is determined by a word's lexical properties, which are catalogued in its lexical entry. The second factor, however, goes beyond detailed lexical properties and involves a more abstract repertoire of form-meaning-metrics units, which form the larger patterns that organize the oral poetic performance. In the general language domain, these

23 The complex relationship between different frames is accounted for by an elaborate hierarchy of frames that stand in different relations to each other, such as inheritance, using, subframe, and perspective on. The FrameNet frame grapher (https://framenet.icsi.berkeley.edu/fndrupal/FrameGrapher) is a visual representation tool that shows the various relationships between frames.

more general properties are captured by grammatical descriptions and analyses. Poetry, however, has a much more restrictive set of "patterns", which are often very formulaic and idiomatic. In the following paragraphs, I first present a particular linguistic theory, namely Construction Grammar (CxG), which, in my view, is well suited for the analysis of oral poetry, in particular formulas. After reviewing some of its basic concepts and ideas, I will show how CxG can be applied to the study of idiomatic and formulaic language found in oral poetry.

The main idea behind CxG is that a linguistic model should be able to account for all facets of a speaker's knowledge about language. Unlike many other theories of grammar, CxG does not make any theoretical distinctions between different areas of grammar such as core and periphery[24] and therefore aims to achieve full coverage of the relevant facts of a language.[25] CxG does not limit itself to analyzing a slice of interesting data representing regular processes in the grammar of a language (e.g., subject-predicate constructions, relative clause constructions, wh-question constructions). Rather, CxG is also concerned with accounting for semi-productive processes (e.g., *way-constructions*,[26] *What's X doing Y?*,[27] *Ditransitive*,[28] *let alone*,[29] and idiomatic constructions[30] (e.g., *kick the bucket; the Xer, the Yer*)) by using the same kinds of principles used to account for more regular processes.

Rather than positing distinct modules of grammar that interact with each other, CxG takes the notion of the linguistic sign[31] as central and posits that each particular form is associated with a specific meaning that licenses as well as constrains it.[32] Goldberg gives the following definition of a construction.

---

24 cf. Noam Chomsky, *Lectures in Government and Binding*. Dordrecht 1981.

25 This section is based on Hans C. Boas, *A constructional approach to resultatives*. Stanford: 2003.

26 Adele Goldberg, *Constructions*. Chicago 1995.

27 Paul Kay/Charles Fillmore, "Grammatical constructions and linguistic generalizations: The 'What's X doing Y?' Construction". *Language* 75 (1999): 1–33.

28 Goldberg, *Constructions*; Hans C. Boas, "Comparing constructions across languages". In: H. C. Boas (ed.), *Contrastive Studies in Construction Grammar*. Amsterdam and Philadelphia 2010, 1–20.

29 Charles J. Fillmore, Paul Kay, and Mary O'Connor, "Regularity and idiomaticity in grammatical constructions: The case of 'let alone'". *Language*, 64 (1988): 501–538.

30 Stefanie Wulff, "Words and idioms". In: T. Hoffmann and G. Trousdale (eds.), *The Oxford Handbook of Construction Grammar*, 274–289. Oxford 2013.

31 Ferdinand de Saussure, *Cours de linguistique générale*. Lausanne 1916.

32 There exist various versions of CxG, but all constructionists subscribe to the basic idea that language consists of a structured inventory of constructions. While some constructionists are interested in typological insights, others focus on the psychological motivation behind language, while others are interested in developing more formalized versions of constructional analyses that

> C is a CONSTRUCTION iff_{def} C is a form-meaning pair $\langle F_i, S_i \rangle$ such that some aspect of F_i or some aspect of S_i is not strictly predictable from C's component parts or from other previously established constructions.[33]

The relationship between form and meaning/function is illustrated in Figure 3, in which the entire box represents the concept of a construction, in which the form (the box on the top) is linked to the meaning (the box on the bottom) via a symbolic correspondence link.

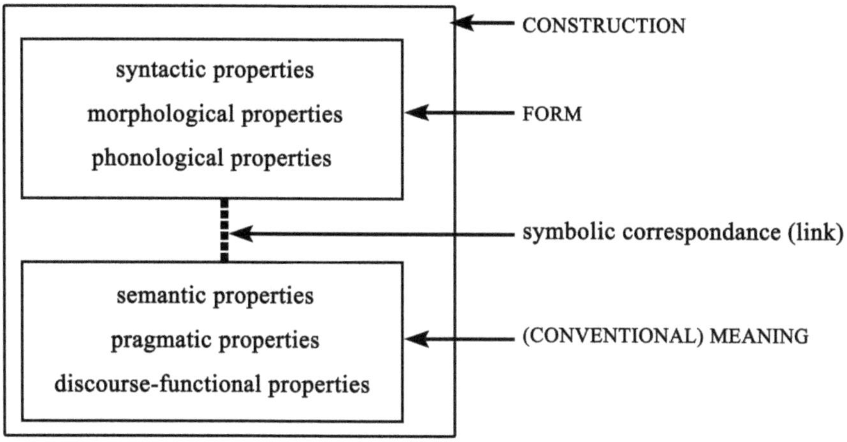

Figure 3: Relation between form and meaning in a construction[34]

The form of a construction can be associated with different kinds of grammatically relevant information that can be semantic, pragmatic, syntactic, morphological, phonological, or lexical in nature. Taken together, language is regarded as a structured inventory of constructions of various degrees of schematicity, ranging from specific morphemes and words to idioms and more schematic constructions such as the ditransitive and the passive, as the following table illustrates.[35]

---

can be used for computational applications. For more details, see the various contributions in Thomas Hoffmann and Graeme Trousdale (eds.), *The Oxford Handbook of Construction Grammar*, Oxford 2013.

33 Goldberg, *Constructions*, 4.

34 William Croft, *Radical Construction Grammar*. Oxford 2001, 18.

35 Unlike many other theories, CxG claims that each language consists of its own inventory of constructions, and that there are many language-specific constructions (as opposed to "univer-

Table 1: Examples of constructions with different degrees of complexity/schematicity[36]

Morpheme	e.g. *pre-, -ing*
Word	e.g. *avocado, anaconda, and*
Complex word	e.g. *daredevil, shoo-in*
Complex word (partially filled)	e.g. [N-s] (for regular plurals)
Idiom (filled)	e.g. *going great guns, give the Devil his due*
Idiom (partially filled)	e.g. *jog* <someone's> *memory, send* <someone> *to the cleaners*
Covariational Conditional	The Xer the Yer (e.g. *the more you think about it, the less you understand*)
Ditransitive (double object)	Subj V Obj_1 Obj_2 (e.g. *he gave her a fish taco; he baked her a muffin*)
Passive	Subj aux VP_{PP} (PP_{PP}) (e.g. *the armadillo was hit by a car*)

The representations of the constructions in Table 1 are not complete as they only represent the most important aspects of the form of the respective constructions, but not the meaning side, or other constructional properties, such as specifications regarding which other constructions they can combine with and the types of restrictions that constrain such combinations. For example, a complete constructional representation of the morpheme *pre-* in Table 1 would consist not only of the form side (in this case three English sounds ordered in a specific order), but it would also come with a description of the meaning side about what *pre-* actually means (represented in terms of a semantic frame specific to time and ordering of events), together with specifications about what other types of constructions it can combine with (in this case specific words such as particular nouns, this is the open slot to the right of *pre-*), and a set of constraints that restrict such combinations so that the inventory of constructions does not license any unattested forms. Kay and Fillmore put it as follows: "[T]he construction grammarian is required to develop an explicit system of representation, capable of encoding economically and without loss of generalization all the constructions (or patterns) of the language, from the most idiomatic to the most general."[37]

sal" constructions). See Croft, *Radical Construction Grammar*, and Boas, "Comparing constructions".

36 Adele Goldberg, *Constructions at Work*. Oxford 2006, 5.

37 Kay and Fillmore, "Grammatical constructions", 2. For a formal implementation of constructional representation, see Charles J. Fillmore, Russell Lee-Goldman, and Russell Rhomieux, "The FrameNet-Constructicon". In: Hans C. Boas and Ivan Sag (eds.), *Sign-Based Construction Gram-*

Other constructions such as partially filled idioms (*send someone to the cleaners*) or covariational conditionals (*the Xer the Yer*) are more abstract, because their open slots (*someone, Xer – Yer*) exhibit fewer restrictions about what types of elements can fill these slots. While the ditransitive and the passive are even more abstract/schematic than the other constructions in Table 1 (the constraints on the elements filling their slots are even broader), there are even more abstract constructions such as relative clauses and subject-predicate agreement constructions.

The idea that language consists of a structured inventory of constructions means that for an utterance to be acceptable in a language, it needs to be licensed by one or more constructions. Consider, for example, a sentence like *A dozen roses, Nina sent her mother!*, which, according to Goldberg[38] is licensed by the following constructions: the ditransitive construction, the topicalization construction, the verb phrase construction, the noun phrase construction, the indefinite determiner construction, the plural construction, and the various constructions representing the individual words (*dozen, rose, Nina, send, mother*). Each of these constructions belong to our inventory of constructions.

One advantage of this approach is that it also allows us to systematically deal with idiomatic and non-idiomatic language use alike, as well as different interpretations of the same utterance caused by two competing sets of constructions. Consider a sentence such as *What is that fly doing in my soup?*, which can have two interpretations, according to Kay and Fillmore.[39] The first one is the literal interpretation, which is licensed by a general *wh*-construction, the noun phrase construction, the prepositional phrase construction, the progressive aspect construction, and the constructions representing the individual words. A compositional interpretation of the meanings of all the individual constructions involved in licensing such a question could result in the waiter answering *I believe that's the backstroke*. The second, and perhaps more likely interpretation, according to Kay and Fillmore,[40] is licensed by a particular sentence level construction with a "specific semantic interpretation associated by convention with just such sentences, interpretations that are neither given by ordinary compositional processes nor derived from a literal meaning by processes of conventional reasoning." This specific incongruous interpretation of the question is thus associated with a particular sentence-level idiomatic construction that is conventionally associated

---

mar. Stanford 2012, 283–299; and Ivan Sag, "Sign-based Construction Grammar: An informal synopsis". In: H. C. Boas and I. Sag (eds.), *Sign-based Construction Grammar*. Stanford 2012, 69–202.

38 Goldberg, *Constructions at Work*, 21.

39 Kay and Fillmore, "Grammatical constructions".

40 Ibid., 4.

with this meaning, and whose meaning cannot be derived compositionally. This is just one of many cases of conventationalized idiomatic expressions found in everyday language.[41]

Over the past five years, construction grammarians have begun to look into adopting the same methodology underlying the description and analysis of semantic frames for the description and analysis of grammatical constructions. The idea behind this move is that just like individual words (or LUs) are capable of evoking semantic frames (with open slots to be filled by FEs), so, too, can particular combinations of words evoke specific grammatical constructions (some of which carry specific meanings which in turn can be represented by semantic frames) with open slots to be filled by Construction Elements. This observation has led to a number of case studies applying the frame-semantic methodology underlying the creation of FrameNet to the creation of a so-called Constructicon, an electronic database with a structured inventory of constructions. Similar to the work-flow in FrameNet, electronic corpora are used to discover sentences illustrating particular constructions. These are then annotated with the relevant constructional information, resulting in an inventory of constructions, in which each construction entry consists of a description of its form side, including specifications of the individual parts of the construction (including information such as phrase type, grammatical function, etc.), together with a prose description of its meaning side (i.e. its interpretation) and a set of annotated sentences exemplifying the construction in context. The creation of construction entries follows one of the basic tenets of CxG (see Goldberg's definition of construction above): every time when a particular pattern cannot be accounted for based on combinations of constructions already recognized, then a new construction entry must be posited.[42]

How are insights from CxG useful for studying oral poetry? Just like frame-evoking LUs have FE slots depending on the semantic frame, constructions also have slots and often associated with meaning. When studying oral poetry, researchers are working with texts composed almost completely of idiomatic expres-

---

41 See Hans C. Boas, "Cognitive Construction Grammar". in: T. Hoffmann and G. Trousdale (eds.), *The Oxford Handbook of Construction Grammar.* Oxford 2013, 233–254.

42 For details on the Constructicon, see Charles J. Fillmore, "Border conflicts: FrameNet meets Construction Grammar". *EURALEX XIII Conference.* Barcelona 2008; Fillmore et al., "FrameNet-Constructicon"; Hans C. Boas, "Zur Architektur einer konstruktionsbasierten Grammatik des Deutschen". In: A. Lasch and A. Ziem (eds.), *Grammatik als Inventar von Konstruktionen? Sprachliches Wissen im Fokus der Konstruktionsgrammatik,* Berlin and Boston 2014, 37–63; and Alexander Ziem, Hans C. Boas, and Josef Ruppenhofer, "Grammatische Konstruktionen und semantische Frames für die Textanalyse". In: J. Hagemann and S. Staffelt (eds.), *Syntaxtheorien. Analysen im Vergleich.* Tübingen 2014, 297–333.

sions, or of partially idiomatic expressions that follow very specific schematic patterns. For example, certain types of speeches are always introduced by a fixed set of arranged elements, such as *optional particle + optional qualification phrase* (e.g. "looking at him/her darkly") + *subject + language verb* ("said/replied/answered"). Such semi-idiomatic expressions are nothing else but a type of semi-idiomatic construction that pairs a specific form with a specific meaning, thereby constituting a form-meaning, a construction. Such constructions are re-occurring entities in oral poetry; they are conventionalized, and stored as patterns in memory. Thus, whenever such an array of words occurs together in a text as a formula, the audience will immediately know what it means, the same way it associates specific knowledge with themes when semantic frames are evoked by specific LUs.

3 Application of Frame Semantics and Construction Grammar to oral poetry

Applying the principles of Frame Semantics and CxG to the study of oral poetry means to first assemble an (electronic) corpus of texts. The analysis of the text should first focus on identifying the relevant frame-evoking LUs and then annotating the text according to the relevant frames and their FEs. If the text is in English, the Berkeley FrameNet database can be used as a reference for finding the relevant frames. Note that FrameNet has not yet covered all of the core vocabulary of English, which means that in cases where there are no existing frames in FrameNet, one would have to make up such frames on the fly.[43] There currently exist FrameNets for other languages, too, such as Japanese, Swedish, German, and Spanish. These lexical databases are built on almost the same set of frames as those of the Berkeley FrameNet for English.[44] For the multitude of languages for which there is currently no FrameNet available, researchers could consult the English FrameNet database in Berkeley to start building an ad-hoc inventory for their own language.[45]

---

43 See Aljoscha Burchardt, Katrin Erk, Anette Frank, Andrea Kowalski, Sebastian Padó, and Manfred Pinkal, "Using FrameNet for the semantic analysis of German: annotation, representation, and automation". In: H. C. Boas (ed.), *Multilingual FrameNets: Methods and Applications.* Berlin and New York 2009, 209–244.

44 See Hans C. Boas (ed.), *Multilingual FrameNets in Computational Lexicography*. Berlin and New York 2009.

45 For various methods, see Miriam R.L. Petruck, "Typological considerations in constructing a Hebrew FrameNet". In: H. C. Boas (ed.), *Multilingual FrameNets: Methods and Applications*. Berlin

To see what an annotation of a text might look like, consider the beginning of the story “The Tiger of San Pedro” by Sir Arthur Conan Doyle.[46] The first sentence reads as follows:

(1) A COLD$_{\text{Ambient_temperature}}$ and melancholy WALK$_{\text{Self_motion}}$ of a COUPLE$_{\text{Cardinal_numbers}}$ of MILES$_{\text{Measure_linear_extent}}$ BROUGHT$_{\text{Bringing}}$ us to a HIGH$_{\text{Dimension}}$ wooden gate , which opened into a gloomy AVENUE$_{\text{Roadways}}$ of chestnuts .

The annotation in sentence (1) shows the frame-evoking LUs in capital letters, followed by the names of the evoked frames in subscript. For example, *cold* evokes the `Ambient_temperature` frame and *walk* evokes the `Self_motion` frame. This first sentence illustrates that there are least seven frames evoked by seven different LUs, but there are other LUs, which have not been annotated, such as *melancholy*, *wooden*, *gate*, *to open*, *gloomy*, and *chestnuts*, which all evoke other frames. Focusing on the seven LUs evoking the seven different frames in (1), we can see how interconnected the different LUs are in this first sentence, i.e. many frame-evoking LUs are themselves FEs of other frames. In other words, there are multiple layers of frames whose meanings are interconnected with each other. Consider the first frame-evoking target LU, which evokes the `Ambient_-temperature` frame in (2).

(2) A COLD^{Tgt} and melancholy [$_{\text{Circumstances}}$ walk of a couple of miles] brought us to a high wooden gate , which opened into a gloomy avenue of chestnuts .[$_{\text{Place}}$ DNI][$_{\text{Time}}$ DNI]

The `Ambient_temperature` frame is defined in FrameNet as the specification of a Temperature in a certain environment, determined by Time and Place. In (2), *cold* evokes the `Ambient_temerature` frame, and the Place and Time FEs are not overtly mentioned (they are definite null instantiations, as can be seen at the end of the sentence). The only overtly realized FE is Circumstances, *walk of a couple of miles*, which serves as some specification of the circumstance under which the ambient temperature is as specified. Note that one LU of the FE Circumstances of the `Ambient_temerature` frame, namely *walk*, is itself a frame-evoking LU, as is shown in the following annotation.

---

and New York 2009, 183–208; Guillaume Pitel, “Cross-lingual labeling of semantic predicates and roles: A low-resource method based on bilingual L(atent) S(emantic) A(nalysis)”. In: H. C. Boas (ed.), *Multilingual FrameNets: Methods and Applications*, Berlin and New York 2009, 245–286; and Schmidt, “Kicktionary”.

46 See https://framenet.icsi.berkeley.edu/fndrupal/index.php?q=fulltextIndex.

(3) A cold and melancholy WALK^{Tgt} [$_{\text{DISTANCE}}$ of a couple of miles] brought [$_{\text{SELF\_MOVER}}$ us] to a high wooden gate, which opened into a gloomy avenue of chestnuts.

In (3), the LU *walk* evokes the `Self_motion` frame, which is defined as the SELF_MOVER, a living being, moves under its own direction along a PATH. Core FEs of the `Self_motion` frame include AREA, DIRECTION, GOAL, PATH, SOURCE, SELF_-MOVER, and SOURCE. In (3), only two FEs of the `Self_motion` frame are realized with *walk*, namely DISTANCE (a couple of miles) and SELF_MOVER (us). Next, consider the annotation of *a couple of miles* in (4), which serves as a part of the CIRCUMSTANCES FE in (2) above. A part of this phrase, *couple*, serves as a frame-evoking target LU evoking the `Cardinal_numbers` frame, of which *of miles* serves as the ENTITY FE.

(4) A cold and melancholy walk of a [$_{\text{NUMBER}}$ COUPLE^{Tgt}] [$_{\text{ENTITY}}$ of miles] brought us to a high wooden gate, which opened into a gloomy avenue of chestnuts.

The LU *miles*, which is a part of the FE ENTITY in (4), is also a frame-evoking LU, which evokes the `Measure_linear_extent` frame in (5). Here, *miles* itself is also the UNIT FE, while *a couple* and *of* are both part of the COUNT FE of the `Measure_linear_extent` frame.

(5) A cold and melancholy walk of [$_{\text{COUNT}}$ a couple] [$_{\text{COUNT}}$ of] [$_{\text{UNIT}}$ MILES^{Tgt}] brought us to a high wooden gate, which opened into a gloomy avenue of chestnuts.

Next, consider the LU *brought* in (6), which evokes the `Bringing` frame. In (6), the AGENT FE of the `Bringing` frame consists of the entire noun phrase *a cold and melancholy walk of a couple of miles*, while *us* is the THEME FE, and *to a high wooden gate, which opened into a gloomy avenue of chestnuts* is the GOAL FE of the `Bringing` frame.

(6) [$_{\text{AGENT}}$ A cold and melancholy walk of a couple of miles] BROUGHT^{Tgt} [$_{\text{THEME}}$ us] [$_{\text{GOAL}}$ to a high wooden gate, which opened into a gloomy avenue of chestnuts.]

In (7), *high* evokes the `Dimension` frame, of which *wooden gate* is the only overtly realized core FE, namely the OBJECT FE. The FE MEASUREMENT is not overtly realized, but is an indefinite null instantiation.

(7) A cold and melancholy walk of a couple of miles brought us to a HIGH^{Tgt} [$_{\text{OBJECT}}$ wooden] [$_{\text{OBJECT}}$ gate], which opened into a gloomy avenue of chestnuts. [$_{\text{MEASUREMENT}}$ INI]

Finally, consider the LU *avenue*, which evokes the `Roadways` frame in (8), of which *avenue* itself is the FE Roadway, and *of chestnuts* the FE Abundant_entities.

(8) A cold and melancholy walk of a couple of miles brought us to a high wooden gate, which opened into a gloomy [$_{\text{Roadway}}$ AVENUE^{Tgt}] [$_{\text{Abundant\_entities}}$ of chestnuts].

The annotation of the first sentence of "The Tiger of San Pedro" shows a number of important points. First, most words in a text are capable of evoking semantic frames whose meanings are part of a language user's linguistic inventory, together with a great deal of world knowledge. Second, any utterance with multiple frame-evoking LUs consists of multiple layers of meaning. Third, FEs belonging to a specific frame can themselves be frame-evoking target LUs, which means that they can serve multiple functions at the same time. Fourth, not all FEs of a frame need to be overtly realized, because they can be inferred from the context, based on the frame evoked by a target LU. This last point is perhaps one of the most intriguing points of a full-text annotation, because one might begin to wonder whether there is any special motivation leading an author or a poet to leave certain FEs unmentioned.

So far, I have only discussed the different semantic frames evoked in one sentence. In doing so, I have given an example of how researchers of oral poetry can in principle apply the same methods to the oral performance of poems. The variation on the different themes evoked by specific words might in this context be of great interest, because it allows researchers to systematically relate different versions of the same poem to each other by identifying the frame-evoking LUs that serve as paraphrases of each other.

One important aspect that I cannot address in detail here concerns the formulaic aspects of full text analysis. Using the analytical machinery of CxG as outlined in the previous section would require the researcher to identify all relevant constructions in (1) (as well as all other sentences of a text), and to carefully formulate constructional entries specifying the form and meaning of each construction. For example, the sentence in (1) is licensed by a number of different constructions, including at least the coordination construction (*X and Y*), the noun phrase construction, the prepositional phrase construction, the relative clause construction, the subject-predicate agreement construction, and all individual words, which represent constructions themselves. Space constraints do unfortunately not allow me to formulate specific construction entries here.[47]

---

47 The interested reader should consult Fillmore et al., "FrameNet-Constructicon"; Boas, "Zur Architektur"; and Ziem et al., "Grammatische Konstruktionen" for details.

While the discussion in this section has focused primarily on applying Frame Semantics and Construction Grammar to the analysis of written texts, its applicability to practices as formulaic and traditional as oral poetry is obvious. First, oral traditions hold on to the sets of encyclopedic knowledge of a community and pass them on across generations. This encyclopedic knowledge is encapsulated in the semantic frames evoked by the key words expressing the main ideas of a poem. As such, semantic frames can be viewed as structuring devices for oral poetry. Even though the oral poet is constantly involved in constructing conventionalized form-meaning templates under narrative, structural, metrical, melodic, and prosodic conditions, he must still have a mental summary of the narrative, a condensed plot that is chronologically organized, and that systematically relates different parts of a story line to other parts. It is exactly these key words that can be analyzed from the perspective of Frame Semantics. In other words, one might hypothesize that oral poets internalize sequences of frames (not specific words) representing different parts of a story line. On different occasions, an oral poet might then use different words to express the same meaning, because these different words evoke the same semantic frame. This is one way that a certain degree of variability arises while still adhering to the basic story line.

For example, a battle that may form the central part of an oral performance has a very specific order of sub-events, each of which can be characterized in terms of semantic frames. First, there need to be opposing parties. The member(s) of these parties could be referred to as *soldiers*, *fighters*, or *mercenaries*, each evoking the same semantic frame that calls for a Frame Element labeled `Party 1` or `Party 2`. Second, at least one party needs to prepare for battle, a process consisting itself of multiple sub-events, each of which can again be characterized in terms of a series of inter-connected semantic frames and the various words which they evoke. Third, the battle itself can be analyzed as a series of sub-events, each characterized by a series of interconnected frames that can be evoked by semantically related words. Finally, a battle typically ends in victory for one party and defeat for another party, each of which can again be characterized in terms of a series of interconnected frames evoked by semantically similar words. This very brief example shows how Frame Semantics can be directly applied to the characterization of story lines, events, and interconnected relations that form the centerpiece of oral traditions.

While semantic frames help with the structuring of oral performances, grammatical constructions help us with understanding some of the key phenomena recurring across oral traditions, such as epic formulaic style and story-patterns. Even though these are natural products of our capacities for meaning construction, they are still limited by the availability of certain types of constructions, spanning the full range from idiomatic to semi-idiomatic to abstract constructions. In other words, oral poets re-construct the details of the story line every time anew, and it is exactly this online, improvisation-based poem-producing techniques that are limited by specific formal requirements of the grammar of a language. By carefully studying the broad variability of epic poems, construction grammarians can investigate and analyze the inventory of grammatical constructions that serve to regulate and produce oral performances. Since grammatical constructions are pairings of meaning with form, they offer a straightforward method for integrating different aspects of form, i.e. not only purely syntactic information, but also information about metrical, melodic, and prosodic conditions in which oral poets are producing their songs.

Frame Semantics and Construction Grammar are not only useful tools for analyzing oral poems, but they can also greatly benefit from the analysis of oral poems to help us understand the nature of spoken language. Since oral poetry is a much more natural form of verbal creativity than written literature, it represents a much more fertile ground of investigation that will help us understand how an elaborate verbal skill is acquired through instance-based generalizations emerging from usage. In other words, a systematic study of large collections of oral poems will surely help cognitive linguists to better understand the many different cognitive processes at work when memorizing, performing, and interpreting oral poetry.

4 Conclusions

In this paper I presented the main ideas and principles of Frame Semantics and Construction Grammar, and I argued that they are useful for the study of oral poetics because they allow researchers to systematically analyze oral poems. This approach goes beyond a close reading of respective passages because it utilizes the toolset of a unified cognitive linguisic approach built on empirical data that allows to cross-check insights from the analysis of poems with other linguistic data. In other words, the analytical toolsets of Frame Semantics and Construction Grammar lend themselves very well to formulaic expressions, particularly the enhanced idiomaticity of oral poetry, which is much stronger in the epic (lyric is much shorter and one can learn stanzas by heart more easily).

Clearly, this paper is only a first step towards applying Frame Semantics and Construction Grammar to oral poetry. To arrive at a more systematic framework for studying oral poetry, a number of steps must be taken. First, researchers will need to agree on specific corpora to be studied. Should these corpora combine transcripts of performances by different poets? Should these corpora span a poet's entire repertoire or only specific aspects? These are important aspects that need to be answered before actually analyzing oral performances. Without a suitable corpus, any usage-based analysis will have severe limitations.[48] Second, to what degree is it possible to re-use existing descriptions of semantic frames and constructions for the analysis of oral poetry? Research on spoken language suggests that while there are significant overlaps in form and meaning between written and spoken language, there are also some significant differences.[49] Third, recent psycholinguistic research has focused on how word meanings and constructions are acquired, used, and processed.[50] Before going into the details of analyzing live performances of poems one should have a clearer understanding of the particular cognitive processing factors involved in such performances. A clearer understanding of such factors will also likely help us with teasing out some of the differences between spoken and written language. Fourth, what is the role of language variation in oral poetry? Do all performances of a particular poem rely on the same standard variety of a language? What if there did not yet exist a standard variety of a language, and poems were written in local varieties? What happens when these poems are performed by poets speaking a different variety of the language? How does that influence the performance itself and how it is received by the audience (who might not speak the same variety of the language)? Research on cognitive aspects of language variation suggests that these are also important aspects that need to be considered when analyzing linguistic data.[51]

Obviously, much future research is required to investigate these questions and many more. The goals of this paper have been more modest: to show that oral composition-in-performance is interesting for linguists and cognitive scientists,

---

48 See Stefan Gries, "Data in Construction Grammar". In: T. Hoffmann and G. Trousdale (eds.), *The Oxford Handbook of Construction Grammar*. Oxford 2013, 93–110.

49 See Susanne Günthner and Wolfgang Imo (eds.), *Konstruktionen in der Interaktion*. Berlin and New York 2007; Arnulf Deppermann, "Konstruktionsgrammatik und Interaktionale Linguistik: Affinitäten, Komplementaritäten und Diskrepanzen". In: A. Lasch and A. Ziem (eds.), *Konstruktionsgrammatik III. Aktuelle Fragen und Lösungsansätze*. Tübingen 2011, 205–238.

50 Giulia Bencini, "Psycholinguistics". In: T. Hoffmann and G. Trousdale (eds.), *The Oxford Handbook of Construction Grammar*. Oxford 2013, 379–396; Friedemann Pulvermüller, Bert Cappelle, and Yury Shtyrov, "Brain basis of meaning, word, constructions, and grammar". In: T. Hoffmann and G. Trousdale (eds.), *The Oxford Handbook of Construction Grammar*. Oxford 2013, 397–418.

51 See Jan-Ola Östman/Graeme Trousdale. "Dialects, discourse, and Construction Grammar", in: T. Hoffmann/G. Trousdale (eds.), *The Oxford Handbook of Construction Grammar*, Oxford 2013, pp. 476–490; and Willem Hollmann, "Constructions in Cognitive Sociolinguistics", in: T. Hoffmann/G. Trousdale (eds.), *The Oxford Handbook of Construction Grammar*, Oxford 2013, pp. 491–510.

and to suggest that Frame Semantics and Construction Grammar together offer a unique set of analytical tools for systematically analyzing oral poetry.

References

Atkins, Beryl T.S. "Analyzing the Verbs of Seeing: A Frame Semantic Approach to Corpus Lexicography". In: Christopher Johnson et al. (eds.), *Proceedings of the Twentieth Annual Meeting of the Berkeley Linguistics Society*, 42–56. Berkeley 1994.

Atkins, Beryl T.S. "The Role of the Example in a Frame Semantics Dictionary." In: Masayoshi Shibatani and Sandra A. Thompson (eds.), *Essays in Semantics and Pragmatics—In Honor of Charles J. Fillmore*, 25–42. Amsterdam and Philadelphia 1995.

Atkins, Beryl T.S. and Michael Rundell. *The Oxford Guide to Practical Lexicography*. Oxford and New York 2008.

Bencini, Giulia. "Psycholinguistics". In: Thomas Hoffmann and Graeme Trousdale (eds.), *The Oxford Handbook of Construction Grammar*, 379–396. Oxford and; New York 2013.

Boas, Hans C. *A constructional approach to resultatives*. Stanford: CSLI Publications 2003.

Boas, Hans C. "From theory to practice: Frame Semantics and the design of FrameNet". In: Stefan Langer and Daniel Schnorbusch (eds.), *Semantik im Lexikon*, 129–160. Tübingen 2005.

Boas, Hans C. "Semantic frames as interlingual representations for multilingual lexical databases". *International Journal of Lexicography* 18.4 (2005): 445–478.

Boas, Hans C. (ed.). *Multilingual FrameNets in Computational Lexicography: Methods and Applications*. Berlin and New York 2009.

Boas, Hans C. "Recent trends in multilingual computational lexicography". In: Hans C. Boas (ed.), *Multilingual FrameNets in Computational Lexicography: Methods and Applications*, 1–36. Berlin and New York 2009.

Boas, Hans C. "Comparing constructions across languages". In: Hans C. Boas (ed.), *Contrastive Studies in Construction Grammar*, 1–20. Amsterdam and Philadelphia 2010.

Boas, Hans C. "Zum Abstraktionsgrad von Resultativkonstruktionen". In: Stefan Engelberg, Kristel Proost, and Anke Holler (eds.), *Sprachliches Wissen zwischen Lexikon und Grammatik*, 37–69. Berlin and New York 2011.

Boas, Hans C. "Cognitive Construction Grammar". In: Thomas Hoffmann and Graeme Trousdale (eds.), *The Oxford Handbook of Construction Grammar*, 233–254. Oxford and New York 2013.

Boas, Hans C. "Wie viel Wissen steckt in Wörterbüchern? Eine frame-semantische Perspektive". *Zeitschrift für Angewandte Linguistik* 57.1 (2013): 75–97.

Boas, Hans C. "Zur Architektur einer konstruktionsbasierten Grammatik des Deutschen". In: Alexander Lasch and Alexander Ziem (eds.), *Grammatik als Inventar von Konstruktionen? Sprachliches Wissen im Fokus der Konstruktionsgrammatik*, 37–63. Berlin and Boston 2014.

Burchardt, Aljoscha, Katrin Erk, Anette Frank, Andrea Kowalski, Sebastian Padó, and Manfred Pinkal. "Using FrameNet for the semantic analysis of German: annotation, representation, and automation". In: Hans C. Boas (ed.), *Multilingual FrameNets in Computational Lexicography: Methods and Applications*, 209–244. Berlin and New York 2009.

Busse, Dietrich. *Frame Semantik: Ein Kompendium*. Berlin and New York 2012.

Chomsky, Noam. *Lectures in Government and Binding*. Dordrecht 1981.

Croft, William. *Radical Construction Grammar: Syntactic Theory in Typological Perspective.* Oxford 2001.
Deppermann, Arnulf. "Konstruktionsgrammatik und Interaktionale Linguistik: Affinitäten, Komplementaritäten und Diskrepanzen". In: Alexander Lasch and Alexander Ziem (eds.), *Konstruktionsgrammatik III. Aktuelle Fragen und Lösungsansätze*, 205–238. Tübingen 2011.
Evans, Vyvyan. *How words mean: Lexical Concepts, Cognitive Models, and Meaning Construction.* Oxford and New York 2009.
Fillmore, Charles J. "Frame Semantics". In: Linguistic Society of Korea (ed.), *Linguistics in the Morning Calm*, 111–137. Seoul 1982
Fillmore, Charles J. "Pragmatically controlled zero anaphora." *Proceedings of the Berkeley Linguistics Society*, 95–107. Berkeley 1986.
Fillmore, Charles J. "Border conflicts: FrameNet meets Construction Grammar." In: Elisenda Bernal and Janet DeCesarius (eds.), *Proceedings of the XIII EURALEX Conference*, 49–68. Barcelona 2008.
Fillmore, Charles J. and Beryl T.S. Atkins. "Toward a Frame-based Lexicon: The Semantics of RISK and its Neighbors". In: Adrienne Lehrer and Eva F. Kittay (eds.), *Frames, Fields and Contrasts: New Essays in Semantic and Lexical Organization*, 75–102. Hillsdale 1992.
Fillmore, Charles J. and Collin Baker. "A frames approach to semantic analysis." In: Bernd Heine and Heiko Narrog (eds.), *The Oxford Handbook of Linguistic Analysis*, 313–340. Oxford and New York 2010.
Fillmore, Charles J., Chris R. Johnson, and Miriam R.L. Petruck. "Background to FrameNet". *International Journal of Lexicography* 16.3 (2003): 235–251.
Fillmore, Charles J., Miriam R.L. Petruck, Josef Ruppenhofer, and Abby Wright. "FrameNet in Action: The Case of Attaching". *International Journal of Lexicography* 16.3 (2003): 297–333.
Fillmore, Charles J., Paul Kay, and Mary O'Connor. "Regularity and idiomaticity in grammatical constructions: The case of 'let alone'". *Language* 64.3 (1988): 501–538.
Fillmore, Charles J., Russell Lee-Goldman, and Russell Rhomieux. "The FrameNet-Constructicon". In: Hans C. Boas and Ivan Sag (eds.), *Sign-Based Construction Grammar*, 283–299. Stanford 2012.
Geeraerts, Dirk. *Theories of Lexical Semantics*. Oxford and New York 2010.
Goddard, Cliff. *Semantic Analysis*. Oxford and New York 2011.
Goldberg, Adele E. *Constructions: A Construction Grammar Approach to Argument Structure.* Chicago 1995.
Goldberg, Adele E. *Constructions at Work: The Nature of Generalization in Language*. Oxford and New York 2006.
Gries, Stefan. "Data in Construction Grammar". In: Thomas Hoffmann and Graeme Trousdale (eds.), *The Oxford Handbook of Construction Grammar*, 93–110. Oxford and New York 2013.
Günthner, Susanne and Wolfgang Imo (eds.). *Konstruktionen in der Interaktion*. Berlin and New York 2007.
Hoffmann, Thomas and Graeme Trousdale (eds.). *The Oxford Handbook of Construction Grammar*. Oxford and New York 2013.
Hollmann, Willem. "Constructions in Cognitive Sociolinguistics". In: Thomas Hoffmann and Graeme Trousdale (eds.), *The Oxford Handbook of Construction Grammar*, 491–510. Oxford and New York 2013.
Kay, Paul and Charles Fillmore. "Grammatical constructions and linguistic generalizations: The 'What's X doing Y?' Construction". *Language* 75.1 (1999): 1–33.

Östman, Jan-Ola and Graeme Trousdale. “Dialects, discourse, and Construction Grammar”. In: Thomas Hoffmann and Graeme Trousdale (eds.), *The Oxford Handbook of Construction Grammar*, 476–490. Oxford and New York 2013.

Petruck, Miriam R.L. “Frame Semantics.” In: Jan-Ola Östman, Jef Verschueren, Jan Blommaert, and Chris Bulcaen (eds.), *Handbook of Pragmatics*, 1–13. Amsterdam 1996.

Petruck, Miriam R.L. “Typological considerations in constructing a Hebrew FrameNet”. In: Hans C. Boas (ed.), *Multilingual FrameNets: Methods and Applications*, 183–208. Berlin and New York 2009.

Petruck, Miriam R.L. and Hans C. Boas. “All in a Day’s Week”. In: Eva Hajicova, Anna Kotesovcova and Jiri Mirovsky (eds.), *Proceedings of CIL 17*. Prague 2003.

Pitel, Guillaume. “Cross-lingual labeling of semantic predicates and roles: A low-resource method based on bilingual L(atent) S(emantic) A(nalysis)”. In: Hans C. Boas (ed.), *Multilingual FrameNets: Methods and Applications*, 245–286. Berlin and New York 2009.

Pulvermüller, Friedemann, Bert Cappelle, and Yury Shtyrov. “Brain basis of meaning, word, constructions, and grammar”. In: Thomas Hoffmann and Graeme Trousdale (eds.), *The Oxford Handbook of Construction Grammar*, 397–418. Oxford and New York 2013.

Ruppenhofer, Josef, Michael Ellsworth, Miriam R. L. Petruck, Chris Johnson, and Jan Scheffczyk. *FrameNet II: Extended theory and practice*. 2010. [http://framenet.icsi.berkeley.edu]

Sag, Ivan. “Sign-based Construction Grammar: An informal synopsis”. In: Hans C. Boas and Ivan Sag (eds.), *Sign-Based Construction Grammar*, 69–202. Stanford 2012.

Saussure, Ferdinand de. *Cours de linguistique générale*. Ed. by Charles Bally, Albert Sechehaye, and Tullio De Mauro. Paris 1916.

Schmidt, Thomas.“The Kicktionary—A multilingual lexical resource of football language”. In: Hans C. Boas (ed.), *Multilingual FrameNets: Methods and Applications*, 101–134. Berlin and New York 2009.

Van Valin, Jr., Robert D. and David P. Wilkins. “The Case for ‘Effector’: Case Roles, Agents, and Agency Revisited”. In: Masayoshi Shibatani and Sandra A. Thompson (eds.), *Grammatical Constructions*, 289–322. Oxford and New York 1996.

Wulff, Stefanie. “Words and idioms”. In: Thomas Hoffmann and Graeme Trousdale (eds.), *The Oxford Handbook of Construction Grammar*, 274–289. Oxford and New York 2013.

Ziem, Alexander. *Frames und sprachliches Wissen: Kognitive Aspekte der semantischen Kompetenz*. Berlin and New York 2008.

Ziem, Alexander, Hans C. Boas, and Josef Ruppenhofer. “Grammatische Konstruktionen und semantische Frames für die Textanalyse”. In: Jörg Hagemann and Sven Staffelt (eds.), *Syntaxtheorien. Analysen im Vergleich*, 297–333. Tübingen 2014.

Anna Bonifazi (Heidelberg)

Particles as cues to structuring in Serbocroatian and early Greek epic[1]

Abstract: This work analyzes analogous functions of particles in two oral epic traditions, and it identifies their overall role as cues to structuring. "Structuring" is a concept adopted from the ethnopoetic studies of Sherzer; it regards the reality of oral performances in the making. Singers use particles to mark the articulation of the narration, and to enact metacommunication—telling about the telling; singing about the singing. Structuring describes a piecemeal progress, and indeed singers tend to use particles in connection to individual steps. However, not only particles but also other discourse markers as well as nonlinguistic devices—attested in the guslar tradition, and to be assumed for the Homeric tradition—are cues to structuring. Nonlinguistic as well as linguistic strategies are used to let composers and receivers follow what is going on in the performance. Therefore, structuring is multimodal. A first implication of the latter point is that particles, far from being unimportant, are precious features indexing a semiotically complex event. A second implication is that epic grammar is to be seen as one of the components of multimodal communication, rather than an independent code. This study can be relevant to research in other oral epic traditions, to research in oral genres beyond epic, and to research in everyday multimodal storytelling from a cognitive and pragmatic point of view.

1 Introduction

This paper reflects work in progress,[2] and therefore presents results that are partial, at least in terms of comprehensive statistics. Nonetheless, it springs from, and pursues, major methodological and theoretical interests. The general aim is to stimulate more research on how the medium of performance influences and

---

1 I wish to thank Cristóbal Pagán Canovás and Mihailo Antovič for organizing a very stimulating and groundbreaking conference. Thanks also to David F. Elmer and Mark de Kreij for their comments to earlier versions of this paper, and to the referees, for helpful criticisms.

2 The analyses and thoughts I am going to offer rest upon two current joint projects, one on the performance of the Serbocroatian song in question, in collaboration with David F. Elmer, and another one on ancient Greek particles in early epic, lyric, drama, and historiography, in collaboration with Annemieke Drummen and Mark de Kreij.

informs the grammar of oral epics, by taking Serbocroatian and early Greek traditions as case studies.

A couple of challenges underlie the comparison of the material and the focus of attention. First, macro asymmetries concern the corpora considered. For the purposes of this paper, guslar tradition ("guslar" is a name applied to Slavic-speaking singers of the Balkans who perform traditional epic songs to the accompaniment of the "gusle", a single-stringed instrument) is represented by a 1000-line song recorded by Milman Parry in Novi Pazar in 1934, titled "Halil Hrnjičić and Miloš the Highwayman" (PN662),[3] which tells about the abduction and rescue of the sister of two heroic brothers, and involves a protagonist, a few helpers, and a villain from the side of the enemies. What represents the early Greek epic tradition is conversely two much longer poems, the *Iliad* and the *Odyssey* (about 27.800 lines altogether) transmitted to us as the masterpieces of the Homeric epic tradition. They center on the battles and the journeys of Achilles and Odysseus respectively, but accounts of numerous other battles, journeys, and events abound. The different size obviously affects the quantity of lexical items I am discussing in this paper, and, more in general, the relationship between the discourse strategies and the different scale of complexity of the stories (plot, network of characters). Beside the variation in length, an even more important asymmetry is to be underscored, that is, the evidence that can be analyzed. We have access to an audio recording in the case of PN662, but in the case of the *Iliad* and *Odyssey* we must content ourselves with texts.[4] Such asymmetries, I posit, can be considered a resource rather than a problem. On the one hand, the study of structure can only be enriched by the consideration of different scales of subject matters and their articulation. On the other hand, the lack of video recordings for both, and of audio recordings for the Homeric part, reminds us of an essential gap underlying both cases: what we study is what is left of communicative events that originally were multimodal (more on that in sections 4 and 5), and therefore complex, regardless of how simple we judge the arrangement of words.

---

3 The original aluminium discs of the recordings of South Slavic epic songs are kept in the Milman Parry Collection of Oral Literature at Harvard University. Several songs have been digitalized and are available online, together with the original transcriptions of the texts by Parry's assistant Nikola Vujnović. The song in question, PN662 (= Parry Number 662) is one of them.

4 A more precise term would be "textualization," for Homeric epic. Ong reminds us (1984:5) that textualization is the process linking orality to writing; it starts "when someone devises a way of putting the words of a language into a script."

A second challenge resides in talking about particles. Even though they do not constitute a fix group of lexical items, particles, or, better to say, discourse particles, are "uninflected words that are only loosely integrated into the sentence structure, if at all" (Fisher 2006:4). They are not a significant part of poetic language, one could object, because they usually do not modify content, and their utterance is usually not associated with poetic styles. Yet, I argue, they reveal something relevant to the presentation of epic content because they represent linguistic cues to the "how" of epic performance, in Serbocroatian decasyllables as well as in early Greek hexameters. The fact that they tend not to modify content does not mean that they are not significant; it only means that their significance lies elsewhere.

The central concept of the article is the idea of "structuring." What is the difference between "structure" and "structuring"? The anthropologist Joel Sherzer extensively studied the oral tradition of the Kuna Indians of Panama, who use neither meter nor rhymes to mark lines and verses, and yet employ a number of nonlinguistic features to "chunk" the text (see in particular Sherzer 1982, and also Hymes 1980 and Woodbury 1985). It is Sherzer (1982: 389) who coins the term "structuring" as opposed to "structure", in order to describe Kuna discourse as the result of an online, dynamic process of composition and articulation. I align with Sherzer in thinking that the idea of structuring (from now on without quotation marks) is more fruitful than that of structure, with regard to oral performances. A methodological benefit is that it discourages the application of pre-determined criteria to the analysis of the shape, size, and quality of possible segments—which could preclude critical findings. A theoretical benefit is that it promotes a route-like (or hodological) rather than map-like (or cartographic) account of the unfolding of oral narratives, which seems to fit the cognitive practice of singers in different traditions (see, for instance, Minchin 2008).

What follows explores the relevance of particle use to structuring (section 2), which is the core of the paper. Sections 3 and 4 analyze further linguistic devices as well as nonlinguistic devices that contribute to structuring. All the features under consideration have a metacommunicative value, as they mark the articulation of epic songs in the making. The final section (5) concludes by pointing out the multimodal nature of structuring. Two theoretical and methodological implications emerge, the former about particle studies, the latter about epic grammar in general.

2 The relevance of particles to structuring in both traditions

In Serbocroatian and in ancient Greek, particles are uninflected and mostly monosyllabic words that guide the processing of the utterances or phrases hosting them. They grammatically range from coordinative conjunctions to interjections.[5] In line with what several works indicate, particles can be subsumed under the larger functional category "discourse markers".[6] Let me anticipate here that for the purposes of this paper, I take discourse markers to include particles as well as various kinds of phrases, and even entire clauses that have a metacommunicative function—see later, section 3. When we infer the metacommunicative functions of particles, our interpretation inevitably combines the morphological, syntactical, semantic, and pragmatic input of their surrounding words. In guslar epic particles occur in combination much less frequently than in Homeric epic; nevertheless, they share with their Homeric counterparts salient characteristics: first, they are regularly, although not always, present; second, most of them occur in clause-initial position—and thus have scope over the entire clause; third, they show some flexibility in use, as lexical choices and discourse functions do not necessarily correlate.

The range of particles used in PN662 is fairly representative of the guslar style. The lexical items are a mix of coordinative conjunctions and primary interjections (*a, ah, aj, e, eh, ej, ha, haj, he, hej, i, no, pa, tek, uh*), with the addition of the exclamatory adverb *aman*. They almost exclusively occur in clause-initial position, and their performance often receives melodic emphasis (see below, passage 14). The research on Homeric particles is complicated by several facts: first and foremost, hexametric lines are longer than 10-syllable lines, which is enough to allow for more options in terms of the size of internal metrical and

5 See Goldman 1990 for a survey of particles in Serbocroatian epic; Dedaić and Mišković-Luković 2010 on discourse particles in South Slavic. Several particles in Homer are discussed in Bakker 2005, in Bonifazi 2012 and in Bonifazi, Drummen, De Kreij 2016.

6 The literature on discourse markers is enormous. Here I just recall a few works that specifically address differences between "discourse markers" and "particles" in definition: Lenk 1997 (discourse markers are a subgroup of particles); Jucker and Ziv 1998 (particles are a subgroup of discourse markers); Schourup 1999 ("markers" regards the function, while "particles" regards the form and the syntactic role). Matras 2000 argues that bilingual speakers tend to eliminate class-distinctions about discourse marking, to simplify monitoring and directing operations in interaction.

syntactic sub-units.[7] Second, it is not clear to what extent participial phrases in Homer can be accounted for as separate narrative acts—while acts in PN662 largely coincide with main or sub-clauses displaying finite verbs.[8] Third, combinations or even clusters of particles are almost the default in Homer, whereas in PN662 such clusters are rare. Fourth, the common terminological distinction between "particles" and "interjections" as far as ancient Greek is concerned is disputable—for instance, *ê* (ἦ) is accounted for as a particle, but is best understood as an interjection.

The next passages—(1) to (6)—are representative of recurrent/common usages of particles, which reflect flexibility in terms of interchangeability. All of them are structuring devices in that they signal the beginning of new narrative and performative steps. In (1a) the same lexical item *a*, which in dictionaries is depicted as a connective expressing mild adversativity, introduces two completely different narrative steps: one is the telling of what Miloš does, and the second is Miloš's summoning of the servants.

(1a) a đe viknu svoje hizmećare ↓
a siđite dolje u mutvaku
...how he [Miloš] calls out to his servants:
"Go down to the kitchen... (PN662,617–618)[9]

Analogously, Homeric epic employs *dé* (δέ, which can also convey contrast) to mark the next step in narration, whether in dialogues ("and...they went on", (1b), line 301, uttered by Laertes) or in the main performer's voice ("then Odysseus... answered", (1b), line 302):

(1b) ....ἦ ἔμπορος εἰλήλουθας
νηὸς ἐπ' ἀλλοτρίης, οἱ δ' ἐκβήσαντες ἔβησαν;
τὸν δ' ἀπαμειβόμενος προσέφη πολύμητις Ὀδυσσεύς
...Did you come as a passenger
on a ship of others, and after dropping you they went on?
Then Odysseus of many counsels answered him [Laertes], and said (*Odyssey* 24.300–302)

7 Bakker 1997 demonstrates that a large part of Homeric hexameters consists of two sub-units each representing a separate intonation unit.
8 Some initial research on mismatches between clauses, lines and discourse acts in both traditions is presented in Bonifazi and Elmer 2012a.
9 Translations of PN662 are by D. F. Elmer (based on those by Lord – see Lord 1954); translations of Homer are my own. As for the mark-up of the Serbocriatian text, see below pp. 141–142.

Flexibility can be observed also in opposite cases, that is, whenever the same kind of narrative step is introduced by different particles. This lets us infer a good amount of interchangeability in expressing sequentiality—"and", "and then", "but then", and "so", which appear to be the most recurrent interclausal connectives in storytelling (Segal et al. 1991). (2a) is a micro-collection of five instances of "when"-clauses in PN662 that follow the end of direct speech. The different types of performative discontinuity associated with their beginning are a further sign of flexibility (see section 4 on the dynamic relation between linguistic and nonlinguistic features of the song). It is notable that the verb employed in third person (referring to hearing or understanding messages) semantically and pragmatically contributes to stress the shift from speech to narration.

(2a)	32 *e kad zaću*	peak + curve	"so, when [the sirdar] heard"
	84 *a kad zaću*	curve	"but when [Tale] heard"
	338 *eh kad zaću*	falsetto	"hey, when [Halil] heard"
	452 *e kad sabra*	-	"well, when [the maiden] understood"
	534 *uh kad sabra*	falsetto	"oh, when [Halil] understood"

The *Iliad* and the *Odyssey* offer different particle choices (*ára, mén, ê*) in the narrative act directly occurring after direct speech, which regularly includes a *verbum dicendi* and a demonstrative form (*hṓs*, ὥς) looking backward to the reported words. Different particles occur also in the immediately following narrative act (*dé, kaí*), which tells about actions and reactions taking place as soon as the speech is over. Here are four representative lines—all of them formulaic:

(2b)	*Iliad* 24.746	Ὣς ἔφατο κλαίουσ', ἐπὶ δὲ στενάχοντο γυναῖκες. "So she [Andromache] spoke while weeping, and the women were adding their lament"
	Iliad 23. 212	Ἣ μὲν ἄρ' ὣς εἰποῦσ' ἀπεβήσετο, τοὶ δ' ὀρέοντο "And she [Iris], having spoken thus, went off, and they [the winds] rose up"
	Odyssey 17.57	ὣς ἄρ' ἐφώνησεν, τῇ δ' ἄπτερος ἔπλετο μῦθος "So he [Telemachus] spoke, and she [Penelope] had no winged words"
	Odyssey 22.465	ὣς ἄρ' ἔφη, καὶ πεῖσμα νεὸς κυανοπρῴροιο "So he [Telemachus] spoke, and the cable of the ship of the dark blue prow..."

Before moving on to more kinds of narrative steps that are very common in both traditions, I recall the notion of discourse acts (De Kreij, this volume). Each narrative step is a strategic unit by means of which a certain communicative action is carried out. As Austin puts it (1962: 8), the uttering of a word "is (...) usually a, or even the, leading incident in the performance of the act (...) but it is far from being usually (...) the sole thing necessary if the act is to be deemed to have been performed". Each action exceeds the mere locution of words. The introduction of new or different discourse acts can be linguistically encoded by dedicated discourse particles, which then refer to the communicative action underlying the utterance. For example, a storyteller may conclude a reported speech by saying: "Well, when the king heard that..." The discourse act performed by "when the king heard that" is to set the next context within which action will take place, and "well" is the discourse marker that introduces at least that act (if not the following as well).

In light of such a consideration, the consistent (albeit not ubiquitous) occurrence of particles at line starts, together with the performative emphasis line starts receive in PN662 (again, consistently but not ubiquitously) makes me connect single lines with single discourse acts. Particles qualify lines as equally-weighted narrative steps, independently of the semantic load—for example, whether a connective is more or less "adversative" and independently of the syntactical role —adverbial, coordinating, or subordinating connective (Bonifazi and Elmer 2012a: 96–99).

The phenomenon of the so-called apodotic occurrences of particles supports the same idea, that is, each line represents an equally important discourse act, whether it is a subclause or a main clause. Even though Homeric lines display a much more variegated picture in terms of matches with clauses, in both traditions main clauses following conditional (whence the term "apodotic", to indicate the occurrence in apodoses), temporal, and clausal clauses may be marked by a correlative particle. *Tek* in Serbocroatian (PN662, 566, 636, 823, 847, 936), and *kaí, allá, ára, dé,* and *dḗ* (καί, ἀλλά, ἄρα, δέ, and δή) in ancient Greek are instances of particles that may occur in apodotic position. (3a) shows an instance of apodotic *teke*:

(3a) ta man behu u tom razgovoru
teke rana zora udarila
Just when they were speaking,
just then the dawn's first light appeared, (PN662,110–111)

(3b) ἀλλ' ὅτε δὴ κοίλη νηῦς ἤχθετο τοῖσι νέεσθαι,
καὶ τότ' ἄρ' ἄγγελον ἧκαν,...
But when the hollow ship was laden for them [the Phoenicians] to go, just then they sent a messenger... (*Odyssey* 15.457–458)

In (3b) the correlation is further underscored by *hóte...tóte* (ὅτε...τότε, "the moment in which"..."at that moment"), which in Homer recurs. One could describe the phenomenon by saying that particles—especially the apodotic ones, which are syntactically superfluous—acquire relevance in pragmatic and performative terms. Particles mark discourse acts, and discourse acts are the fundamental building blocks of the poem. This is where structuring lies: by introducing or marking discourse acts particles reveal the chunking process of a story that has to be received piece by piece. Subclauses and main clauses are alike, to the extent that each of them overlaps with a discourse act and a melodic unit. In making this point I take into account the plentiful recurrence of particles after subordinating conjunctions, which the Homeric poems show, but is hardly notable in PN662.

The next passages illustrate discourse acts that most typically characterize the epic songs. Some particles seem to work better than others in introducing them. The first couple exemplifies the act of listing. Listing implies that a series of elements is presented as sharing some similar pattern (e.g. the same character performing discrete actions to effect an overall move; a synchronically shared situation involving the presence of discrete inviduals or groups; a diachronically shared situation involving a succession of discrete symbolic actions). Catalogues and genealogies constitute ideal frameworks for acts of listing. Structuring in this case results from the pattern of sequencing that is foreshadowed by the utterance of the first entry. (4a) and (4b) feature *e*, *a* (for Serbocroatian), *mén* reinforced by *prôta*, "first", and *aû* reinforced by *deuteron*, "second" (for ancient Greek) in arming scenes.

(4a) *ja poskoči pa se opremijo*
a pripasa silah i oruže
e dofati brešku granajliju
I jumped up and armed myself,
and I put on my belt and weapons,
and I picked up my engraved rifle (PN662,228–230)

(4b) κνημῖδας μὲν πρῶτα περὶ κνήμῃσιν ἔθηκε
καλάς, ἀργυρέοισιν ἐπισφυρίοις ἀραρυίας·
δεύτερον αὖ θώρηκα περὶ στήθεσσιν ἔδυνεν
First, he [Alexander] set the greaves about the legs,
beautiful, fitted with silver leg-guards;
second, he wore the corselet about his chest (*Iliad* 3.330–332)

Another frequent discursive procedure involves shifts from a general indication to a detail attached to that indication. This is shared by narrative accounts as well as by speeches. The first two examples show the employment of *a* and *kaí*, respectively, to encode such a shift.

(5a) kad mu bliže pa se primaćije
a na krilo dupe naticu(je)
When she moves closer to him
and sets her bottom in his lap, (PN662,484–485)

(5b) ἀλλ' ἄγε μοι τόδε εἰπὲ καὶ ἀτρεκέως κατάλεξον
But tell me this, [that is], give me an exact account. (*Iliad* 24.380)

In (5a) the singer uses *a* to introduce a detail in narration. In (5b) the speaker, Hermes, further specifies what he is asking for (the entire line is formulaic, which means that the shift in discourse is formulaic as well). While the translation "and" does not do justice to what the particle *kaí* accomplishes, the metacommunicative meaning of *a* is equivalent to "let me detail what I'm saying". Unlike the Serbocroatian *a*, the Greek *kaí* starts the second half of the metrical line, which coincides with a *kôlon* (I will resume the term towards the end of the current section).

A similar shift can be ascribed to "camera techniques" (Bakker 2005; Minchin 2001; Bonifazi 2008 and 2012; Elmer 2009; Bonifazi and Elmer forthcoming): it is as if the singer in the structuring of the song moves his "camera-in-the-mind" to redirect the visual focus by means of *ad hoc* narrative discourse acts, with close-up and zooming-in effects. In those cases metacommunication relies on visualization.

(6a) e bane mu zdravljom odvaljuje
a do njega Miloš ćesedžija
and the ban replies to him with salutation.
Next to him [was] Miloš the Highwayman, (PN662,597–598)

(6b) ...τὼ δέ οἱ ὤμω
κυρτὼ ἐπὶ στῆθος συνοχωκότε· αὐτὰρ ὕπερθε
φοξὸς ἔην κεφαλήν,...
...and his [Thersites'] shoulders
were humped and contracted over his chest, and above them
his head was distort,... (*Iliad* 2.217–219)

In (6a) *a* suggests a visual shift from the ban to somebody else; analogously, in (6b) *autár* (αὐτάρ) prompts the audience to shift their mind's eye to another side of the picture, in this case another side of the peculiar body of Thersites, whose ugliness in the *Iliad* matches a disruptive character (Elmer 2013: 93–97).

The analysis now moves on to particles that seem to be designed for specialized functions. For example, some particles carry a distinct interpersonal value, and are used to convey special emotional attitudes.

(7a) aman! što se Mujo uplašij(o)
Aman! How Mujo was frightened! (PN662,82)

Aman occurs five times over the song; in four of them it conveys the performer's distress in realizing/commenting on a certain fact. In this case the entire line is an exclamatory act revealing emotional stancetaking. Below, (7b) shows the employment of ancient Greek *dḗ* (δή), a particle that is used in evaluative utterances in other authors as well (such as Herodotus and Thucydides), in the construction *hoîon dḗ* (*hoîon* literally means "what sort of" and is used in interrogative as well as exclamatory clauses). The speaker is the nymph Calypso, who comments on a speech that Odysseus has just delivered.

(7b) ἦ δὴ ἀλιτρός γ' ἐσσὶ καὶ οὐκ ἀποφώλια εἰδώς,
οἷον δὴ τὸν μῦθον ἐπεφράσθης ἀγορεῦσαι.
You definitely are a rogue, and what you know is not pointless;
what a speech you devised to perform! (*Odyssey* 5.182–183)

Unlike Serbocroatian *aman*, ancient Greek *dḗ* is used in epic literature to fulfil other discourse functions as well. Most of all, while (7a) is uttered by the singer, (7b) is uttered by a character.[10] So, this pair of examples shows significant asymmetries. As for Homer, it is quite hard to pin down when and how particles may signal such metacommunicative acts, especially by the primary narrator. More research should be done on Homeric particles in connection with stancetaking and evaluation by the performer; I believe that their metacommunicative potential in this sense is still underestimated.

These asymmetries notwithstanding, a couple of general observations on particles marking stancetaking can be made. Besides the organizational structuring that divide individual narrative steps, particles may suggest emotional structuring, which flags moment of particular personal involvement by a character, or emotional alignment between performer and character. Furthermore, particles in the Homeric epic, and particles or interjections in the guslar epic, make the emotional and the organizational structuring look alike.

The second special design I discuss is the occurrence of particles in non-initial position, that is, at the opening of subsequent *kôla* (*kôlon* in Greek means "member", and it designates a metrical unit that is part of a line). The *deseterac* represents an easy case, as the second *kôlon* invariably starts with the fifth syllable. In PN662 the only particles that appear in that metrical slot are *pa* (36 times), *a* (5 times), and *no* (2 times). Here is an example:

10 I thank Mark de Kreij for signaling that in Homer the combination ἦ μήν, (occurring only two times, in direct speech, at *Iliad* 2.291 and 7.393), resembles *aman* in pragmatic function: it charges the information units it introduces with an exclamatory tone. I would render the respective lines with "ah, there really is toil in a disheartened return!" and "ah, though really the Trojans bid him [Alexander] to do so!"

(8a) a to reće | pa je razminujo
He [the priest] said that, and he departed. (PN662,711)

According to my analysis, most of the times *pa* in that position marks a separate narrative act, as in (8a). As for Homer, several particles and combinations of particles may start a second *kôlon* in hexameters. The latter sometimes coincides with the start of an act that continues in the following line; sometimes it coincides with a full narrative act, as in (8b), introduced by the particle *allá* (ἀλλά):

(8b) Ἥρῃ δ' οὐκ ἔχαδε στῆθος χόλον, | ἀλλὰ προσηύδα
Hera's breast did not contain her wrath, but she spoke: (*Iliad* 4.24)

Once again, more research would be advisable, seeking for the lexical items that are found at the beginning of all the possible internal *kôla* of hexameters, and in connection with discourse acts or part of them.

For the third special design involving epic particles, conversely, a more neat lexical and functional correspondence between Serbocroatian and ancient Greek can be observed. The particles in question, for a change, are mostly used to signal a content-relation. They are *i* and *te* (τε). Consider the following passages:

(9a) *a pripasa silah i oruže*
and I put on my belt and weapons, (PN662,229)

(9b) τόξ' ὤμοισιν ἔχων ἀμφηρεφέα τε φαρέτρην·
[Apollo] with the bow on his shoulders, and the covered quiver (*Iliad* 1.45)

Serbocroatian *i* typically occurs later in the line and may occupy different positions. Non-initial *i* mostly coordinates noun phrases; more importantly, the majority of the combined noun phrases semantically represent an easy or natural association—"fields and meadows", "wine and brandy", "health and well-being". Several times what is coordinated is in fact an hendiadys: "Sultan and Emperor", "glass and window" (for window-glass). (9b) displays a similar function belonging to ancient Greek *te*. Viti (2008) claims that Homeric Greek and Vedic have a coordinator that seems to be specialized in what some linguists call "natural coordination" as opposed to "accidental coordination" (Wälchli 2005) – *te* (τε) and *ca* respectively. Serbocroatian *i*, I submit, shows a similar specialization. In these cases structuring resides in the insertion of pieces of knowledge that are broadly shared by audience and performer, within but also beyond the storyworld.

Let us sum up the contents and the points of this section. Particles are mostly monosyllabic indeclinable words that tend to occur at the beginning of clauses, and signal how to process the host clauses. Passages (1) to (6) have shown them to indicate various typical narrative steps, which coincide with discourse acts. The singer performs acts such as listing entries, zooming in on a detail, and shifting the visual focus of attention. Passages (7) to (9) exemplify acts where emotional stance is conveyed; acts that are shorter than a line; and, finally, particles signalling "natural" semantic associations. All lexical items under consideration are about the process of narration. Within the realm of storytelling, it is part of the narrative dynamic to provide signs about the act of narrating, which Bauman explores in terms of metanarrative signs (1986: 98–101). Therefore, since particles are part of the traditional epic diction, metanarration is as traditional as narration. Moreover, if particles are cues to structuring, structuring is traditional as well.

3 Further linguistic cues to structuring in both traditions

Within language, an indefinite number of expressions/lexical choices may be employed to relate to the context in which communication occurs. Gumperz called them "contextualization cues", as they bridge the gap between what is explicitly said and the knowledge shared by the communication participants on an implicit level. A contextualization cue is "any feature of linguistic form that contributes to the signaling of contextual presuppositions" (Gumperz 1982: 131). The contextualization cues that tell about what is going on in the communication have in fact a metacommunicative function.[11] Particles are contextualization cues whose metacommunicative function consists in showing the structuring of the telling. The point of this section is that further linguistic devices besides particles can perform the same function. A broad label commonly used in linguistics for what encodes cues to the "how" of the message is that of "discourse markers", which I already anticipated at the beginning of section 2:

> Discourse markers are items external to propositional content which are useful in locating the utterance in an interpersonal and interactive dimension, in connecting and structuring phrasal, inter-phrasal, and extra-phrasal elements in discourse, and in marking some on-going cognitive processes and attitudes. (Bazzanella 2006: 456)

11 Maschler (2009: 1) takes Hebrew discourse markers as reflecting metalanguage, that is, language is used to communicate about using language.

To the extent that discourse markers stimulate inferences about the circumstances of discourse that are external to content, they are the linguistic devices that most closely relate to nonlinguistic communication. In what follows, paired examples from the two traditions illustrate various discourse markers that I read as cues to structuring.

The first linguistic cues express communicative immediacy:[12]

(10a) eh Mujo mu bolje prifatijo
sad ovako pa mu govorijo
Hey, Mujo accepted it even more heartily.
Now he [Osman] spoke to him [Mujo] thus. (PN662,171–172)

The hero Osman has just called a greeting to his companion Mujo, and Mujo accepts it heartily. At line 172 *sad*, which means "now", is inserted between the past tense verb *prifatijo* ("accepted") and the past tense verb *govorijo* ("spoke"). The discourse relevance of *sad* is metacommunicative, and is about structuring, once again: *sad* explicitly anchors the singing of the unfolding events to the "here and now" of the unfolding performance. Mujo's acceptance and Osman's speaking are both recalled as having happened in the past and re-enacted as taking place in the present. What follows indeed is Osman's speech re-performed.

The Homeric poems are hardly as objectively explicit as guslar poems in anchoring the heroic events to the present of the performance. However, recent studies in Homeric discourse have revealed a wider range of functions of morpholexical items such as particles and aorist verb forms occurring in the two poems, so as to include very similar pragmatic and metacommunicative functions (Bakker 1993, 2001, 2005; Bonifazi 2012). The following passage may be taken as a case in point:

(10b) Αὐτὰρ ἐπεὶ τό γ' ἄκουσε θεὰ λευκώλενος Ἥρη,
αὐτίκ' ἄρ' Ἥφαιστον προσεφώνεεν ὃν φίλον υἱόν·
So, as the goddess, white-armed Hera, heard this,
right at this moment she spoke to Hephaestus, her dear son: (*Iliad* 21.377–378)

By virtue of its literal meaning "at the same time"/"at the very moment", the ancient Greek adverb *autíka* (αὐτίκα) may be translated as "at that moment" as well as "at this moment", depending on the interpretation of the stance the speaker applies to the narration of what follows, that is, whether it is a distal or

12 For this notion in psycholinguistics, see Chafe 1994: 3; 6; 32; 39 about "immediacy", and 196–198 about "immediate mode of speaking"; for the application of this idea to Homer, see Bakker 2001, and 2005: 94–96; 103–104; 127–135.

proximal stance. The addition of the particle *ára* (ἄρα) after *autíka*, and the fact that the line introduces the re-performance of Hera's speech suggest that the speaker in question (the Homeric performer) expresses no distance towards the narrated events.

The next couple of passages illustrate the idea of communicative immediacy while singers introduce new or already known characters entering the scene. The guslar tradition makes use of special word clusters to do that, which Elmer (2009) calls "presentation formulas". Here is an example from PN662:

(11a) kad evo ti Arnautovića
here you are, the Albanian's son [Osman]. (PN662,150)

Interestingly, the grammar of the formula combines *kad*, which can be taken as a temporal conjunction ("when") but also, alternatively, as a temporal adverb ("at that moment")[13]; *evo*, adverb of place that indicates proximity ("here"); and *ti*, literally "to you". "Here" and "you" convey immediacy.

Even though the *Iliad* and the *Odyssey* lack grammaticalized formulas to introduce the appearance of characters, it can be observed that from time to time the appearance of objects and/or people is announced through the particle combinations *kaì dḗ* (καὶ δή) and *kaì mḗn* (καὶ μήν).

(11b) ὢ πόποι, ἦ μάλα δὴ μετεβούλευσαν θεοὶ ἄλλως
ἀμφ' Ὀδυσῆϊ ἐμεῖο μετ' Αἰθιόπεσσιν ἐόντος·
καὶ δὴ Φαιήκων γαίης σχεδόν, ...
Oh strange! Certainly the gods changed their plan
about Odysseus, while I was among the Ethiopians. Here he is! Near the land of the Phaeacians...(*Odyssey* 5.286–288)

The speaker is the god Poseidon, who is hostile to Odysseus; back from the Ethiopian land, he spots the hero on his way to being rescued, to the god's disappointment (see lines 282–285). *kaì dḗ* is one of the combinations of particles that in later ancient Greek texts, such as tragedies and comedies, do signal the appearance of someone on stage.[14] Both lexical items may convey immediacy even independently of one another, *dḗ* whenever its utterer expresses how clear

---

13 Such an ambivalent role of *kad* is highly significant, as it shows that the quite numerous *kad*-clauses in this type of epic mark single narrative steps independently of the syntactical reading associated with them (as subclauses or as independent clauses). See Bonifazi and Elmer 2012a: 99–100, n.19.

14 E.g. Sophocles, *Ajax* 544 (Tecmessa saying to Ajax: "*kaì dḗ* one of the servants is accompanying him [your son]". See Van Erp Taalman Kip 2009. In *Odyssey* 11, the appearance of Tantalus and

something is appearing to him/her, and *kaí* whenever it marks the instantaneous character of an event. Incidentally, I let readers notice that in the same passage a further combination of three particles, *ê mála dḗ* (ἦ μάλα δή, 286) works as a construction conveying Poseidon's stance towards the plans of other gods about Odysseus.

Another type of cue to structuring that suggests immediacy is a "you" reference that breaks the usual third-person narration. A generic "you"-receiver is evoked while s(he) is potentially watching the scene. Again, the active involvement of an addressee (however unspecified) reflects intended interaction. Metacommunication is in the recalling of an immediate reception of the upcoming piece of performance, which is anticipated to be visually remarkable; in (12a) Miloš, the villain, is going to chase Halil, the main hero, to attack him (900–965); in (12b) Agamemnon is going to move among the ranks, encouraging soldiers individually (220–418).

(12a) vi da vidiš Miloš ćesedžije
You should have seen Miloš the Highwayman, (PN662,900)

(12b) Ἔνθ᾽ οὐκ ἂν βρίζοντα ἴδοις Ἀγαμέμνονα δῖον
Then you would not have seen good Agamemnon slumber (*Iliad* 4.223)

While Serbocroatian evidence is sound on this feature (*vi da vidiš* "you should have seen" and variants are relatively frequent in PN662 as well as in other epic songs), the Homeric counterpart is weaker; such "you"-forms are much less recurrent, and varied by either the negative form (*ouk ídois* "you would not have seen...") or the type of verb ("you would (not) say..."), as in Homer *Iliad* 4.429–30; 5.85–86; 15.697–98; 17.366–67.

Finally, entire clauses may be seen as cues to structuring, as the following questions show:

(13a) e ko beše ćeti ćetobaša
And who was the leader of the raiding party? (PN662,12)

(13b) Τίς τάρ σφωε θεῶν ἔριδι ξυνέηκε μάχεσθαι;
Who, then, of the gods brought these two [Achilles and Agamemnon] to fight? (*Iliad* 1.8)

Sisyphus to Odysseus in the land of the dead is marked by *kaì mḗn*: *Odyssey* 11.582 καὶ μὴν Τάνταλον; 11.593 καὶ μὴν Σίσυφον.

Even though the focus of attention of these interrogative acts regards content, the purpose of performing them is to draw the attention of the audience to the next subject of the ongoing telling. In both cases the next subjects (the character Mujo and the god Apollo) are going to be dominating figures in what follows. The metacommunicative force resides in indirectly asking something about the narrative process ("which character should now enter into the account?", "how should the telling go on now?"). Such questions not only express self-reference, but they also grammaticalize an interactional aspect: the epic performer shows that he negotiates references and meanings by exposing the audience to a potential dialogue. Other instances of metacommunicative questions in PN662 introduce the direct speech of a character ("and what did X say"?). In the *guritan* tradition of South Sumatra (see Collins 1998: 26–38) similar formulaic questions frequently mark points of narrative transition. In these examples it is the entire interrogative clause that works as a discourse marker.

All the discourse markers commented upon in this section create pragmatic vividness. They encode the speaker's immediate perception of a certain reality, which results in the reception "here and now" of the messages: *sad* and *autíka* + *ára* in (10) do so by alluding to the "now" of the re-enactment; *vi da vidiš*, *kad evo ti*, *ouk ídois*, and *kaì dḗ* in (11) and (12) appeal to vision and to a viewing "you"; the questions in (13) involve real or fictional interlocutors. I advance that the "here and now" of epic narration applies to each individual narrative act concerning a certain event, and therefore reflects a piecemeal process, which develops step by step. It is a sign of an online way to proceed; it is a mark of structuring.

4 Nonlinguistic cues to structuring in Serbocroatian epic

Just as the linguistic devices I have discussed so far, several nonlinguistic metacommunicative signs in the guslar tradition are linked to the articulation of songs. These nonlinguistic signs include different melodic patterns, presence vs. absence of rhythmical pulse, falsettos, timbre, tone of voice, manipulation of tempo. Before illustrating how they work in an actual excerpt, it is essential to point out a structural displacement concerning nonlinguistic and linguistic features in PN662.

The displacement in question regards the metrical structure of language and the melodic units of PN662, which is practiced also in other songs by singers of the Novi Pazar area. Each each metrical unit (and therefore each linguistic verse) spans ten syllables – *deseterac*, decasyllabic verse. Each melodic unit spans ten syllables too; however, the melody, which has a contour that usually starts and ends on the same note, ends on the 9th syllable of the decasyllabic verse, followed by a short pause. The new melodic contour starts on the last syllable of the linguistic verse. Therefore there is a one-syllable displacement between the two units. The performative advantage of such a displacement consists in providing continuity on a larger scale, by smoothing what otherwise would be a drastic break (that is, linguistic and non-linguistic at the same time), and by linking melodic and metrical units in a virtually endless chain. Still, the overall effect of the performance is that the singer unfolds the song unit by unit, step by step.

The general significance of this displacement is that nonlinguistic means do not mechanically overlap the linguistic ones. They work together, but they are distinct. According to our research (see in particular Bonifazi and Elmer 2012b), they may reinforce each other, but they also create tensions; furthermore, matches and mismatches are never predictable.

Let me illustrate the dynamic relationship between language and music by looking at an excerpt from PN662. David F. Elmer and I devised a performative transcription of the song that uses a set of simple graphic devices to incorporate relevant features of the performance directly into the transcription of the text. The inevitable approximation of the marking matches the "organic" nature of the singer's vocalizations—each line is similar to, and at the same time it differs from, the others. Besides using the voice, the singer actually plays the gusle (an early account of the music of this tradition is in Herzog 1951); however, the melody of the gusle and that of the voice do not differ enough to justify a separate coding in our mark-up.

(14) **pa evo** je kralju na odaju
sve je redom cura uspavala
kad se oni saćin ućineli
tek dofati sablju Miloševu
a iziđe dolje u avliju
a zađenu sablju u duharu
malo mahnu pa je predvojila
e sad mu balćak u kanu baci—la (851)
obesi je baš đe mu je bila
a dofati dvije sadelije
a odnese dolje u mutvaku
sadelije vinom okvasila
obesi hi baš đe su mu bile
pa dofati brešku granajliju
evo brešku vinom natoćila
obesi je baš đe mu je bila
pa se vrnu dolje kod Haljila

> And here she [Anđelija] is before the King in his chamber,
> the girl put all of them to sleep, one after another.
> When they were sleeping soundly,
> straightaway she took Miloš's sword,
> and went down to the courtyard,
> and thrust the tip of the sword into the wall.
> She twisted it a little and broke it in two.
> And now she put the hilt back in the scabbard, (851)
> she hung it just where it was before,
> and she took his two pistols,
> and brought them down to the kitchen.
> She soaked the pistols with wine,
> she hung them just where they were before,
> and she took his engraved rifle.
> Here now, she filled the rifle with wine,
> she hung it just where it was before,
> and she turned back downstairs to join Halil. (PN662, 844–860)

At this moment of the guslar song, Anđelija, the main female helper of the main hero, Halil, enacts a stratagem in order to disable the enemy (Miloš); she retrieves all of his weapons and makes them useless in various ways. This is a lively excerpt not only because of the narrative speed and the compressed sequence of actions, which almost iconically renders the promptness of Anđelija, but also because of the performative manner. The speed of melody (4,3 seconds per line, while the average in this song is 5 seconds), the frequency of melodic curves including middle-high pitches (represented by light grey areas) and high pitches (represented by dark grey areas), and the very similar pulse of lines 852–854 and 855–860, all of this conveys vividness and suspense. As far as particles are concerned, they almost exclusively occur at the start of lines (see *tek, a, pa,* and *e* in bold).[15] The excerpt displays also *sad* (851) fulfilling the same metacommunicative function as in (10a). Finally, *evo* (858, "here") and *pa evo* (844) work as discourse markers conveying immediacy. Crucially, all these features receive melodic emphasis.

Grey parts tend to occur earlier in the line. They represent melodic pitches that are significantly higher than the base tone. These nonlinguistic features mark the similarity between melodic contours, and shape each of them as a unit. No other objective segmentation above this level is detectable, in fact. An overview of the entire song marked up on written pages (which records the main performative

---

15 Goldman (1990: 219–223) highlights the high frequency of verse-initial connectives. Generally speaking, the first part of the *deseterac* tends to be semantically lighter and syntactically looser (Foley 1995: 19 n 46). In PN662 particles are extrametrical only in exceptional cases; see lines 1 ("a"), 83 ("e"), 589 ("ej") and 591 ("a"). Norrick 2009 explores English primary interjections as marking discourse organization.

events and is naturally arranged *deseterac* by *deseterac*) reveals that there are no major breaks or drastic discontinuities, either linguistically or nonlinguistically.[16] Yet, linguistic and nonlinguistic discontinuities do not necessarily coincide. When changes of character, place, and/or time occur in the narrative, the performance may continue with units closely resembling each other, and, conversely, performative discontinuities may appear when the narrative does not offer any reorientation. For example, the perceptibly slowed down tempo with which the singer performs line 851 definitely breaks the pace of the performance at that moment; however, content does not indicate any particular discontinuity—the words are just "and now she put the hilt back in the scabbard", which relates neither to a climactic point nor to a scene transition.

All of this indicates that the singer uses nonlinguistic as well as linguistic techniques to organize the flow of song. Composers and listeners follow what is going on in the performance thanks to the dynamic combination of multiple codes, thanks to multimodality. Overall they perceive a piecemeal progression of relatively short steps, with new and old information, discontinuous and continuous vocalizations appearing in a balanced alternation. Such a piecemeal movement is what drives the song and guides its structuring.

5 Structuring is multimodal: Two implications

The term "structuring" stresses the online process of attaching step to step, act to act, in order to produce narrative and performative progress. Since structuring involves the use of both linguistic and nonlinguistic devices, structuring is multimodal. Originally, the term multimodality refers to the semiotic complexity of representations involving language-in-use, images, sounds, and gestures on television (Kress and Van Leeuwen 2002). Guslar epic is multimodal insofar as singers use words, different voice modulations, instrumental music, and an extralinguistic setting (a dedicated spatial environment, and traditional costumes) to enact the epic performance.

The multimodality of early Greek epic is supported by a number of direct and indirect sources. They tell us that Homeric epic was enacted through oral performances mingling music and words for centuries and centuries. The specific forms and conditions varied diachronically, of course; the range is from the descriptions

16 From this range we should exclude, however, the very beginning and the very end of the song, which have an exceptional character both in content and performance.

of singers composing while performing within the poems themselves, to a recitative-like declamation of canonical texts. Several nonlinguistic aspects of Homeric poetry are generally speaking acknowledged by scholars, such as the text-internal nonlinguistic behavior of characters, the possible association of some wordings with *ad hoc* gestures, and the cultural significance/effect of singing rather than "plain telling" (see, notably, Bérard 1918). The fact that modern scholarship cannot closely analyze anything but the text of the Homeric poems—the remaining multimodal signs being unattested or too vaguely described—does not justify an a priori disregard for the nonlinguistic components of meaning.

A first implication of epic multimodality concerns the study of particles. The appreciation and understanding of particles in early Greek and in Serbocroatian epic can be achieved only if we assume the centrality of performance. In both traditions they attest to a semiotically complex event. Particles are not fillers or superfluous lexical items to be skipped. Particles, interjections, and further discourse markers *qua* cues to multimodal structuring have to be considered inherent components of the oral epic genre. They should never be omitted in textualization procedures. Moreover, studying them can be relevant to research in oral genres beyond epic, because their use is cross-linguistically widespread (see, e.g. Hymes 1980; Sherzer 1982; Woodbury 1985; Bauman 1986; Collins 1998), and to research in everyday multimodality, because they reflect specific cognitive abilities and pragmatic competence of speakers.

The second implication is about epic grammar in general. In a multimodal approach to communication, the heuristics shifts the focus of attention by decentering the role of language. As Iedema puts it (2003:33):

> In general terms, the trend towards a multimodal appreciation of meaning making centres around two issues: first, the decentring of language as favoured meaning making; and second, the revisiting and blurring of the traditional boundaries between and roles allocated to language, image, page layout, document design, and so on.

The multimodality of guslar epic and of the *Iliad* and the *Odyssey* invites a shift of attention from language to performance. This may lead to a substantial rethinking of the grammatical components of epic diction, at least in the two traditions in question.

The grammar of the text and the nonlinguistic sides of performances are usually considered independent sources of meaning. Scholars talk about metrical fillers and oral syntax, for example, as if grammar constitutes a pre-existing system to be adapted in its syntax and word order to the *constraints* of meter and spokenness. The multimodal domain within which epic performances (at least in these two traditions) are generated and transmitted, however, invites a different view. Rather than being grammar a separate abstract means that is modified to match the idiosyncratic requirements of the epic genre, it is the multimodality of epic communication that shapes the grammar of its linguistic component. This thought aligns with relatively recent general claims about grammar and discourse (e.g. Du Bois 2003:11). Epic singers cannot conceive the production of grammar as separated from the production of meter (on this, see in particular Nagy 2010); the same holds for the production of melodic patterns. Music, meter, and words are generated together and meant to be processed together. This is the second implication: epic grammar (particles included) is better understood as one of the components of epic multimodal genre.

References

Austin, John Longshaw. *How to do things with words.* Cambridge, MA 1962.

Bakker, Egbert J. "Discourse and Performance: Involvement, Visualization, and 'Presence' in Homeric Poetry". *Classical Antiquity* 8 (1993): 5–20.

Bakker, Egbert J. *Poetry in Speech:Orality and Homeric Discourse.* Ithaca and London 1997.

Bakker, Egbert J. "Similes, Augment, and the Language of Immediacy". In: J. Watson (ed.), *Speaking Volumes: Orality and Literacy in the Greek and Roman World*, 1–23. Leiden 2001.

Bakker, Egbert J. *Pointing at the past. From formula to performance in Homeric poetics.* Washington, DC and Cambridge, MA 2005.

Bauman, Richard. *Story, performance, and event. Contextual studies of oral narrative.* Cambridge and London 1986.

Bazzanella, Carla. "Discourse Markers in Italian: Towards a 'Compositional Meaning'". In: Kerstin Fischer (ed.), *Approaches to Discourse Particles*, 449–464. Amsterdam and Boston 2006.

Bérard, Victor. "Le geste de l'aède et le texte homérique". *Revue des Études Grecques* 31 (1918): 1–38.

Bonifazi, Anna. "Memory and Visualization in Homeric Discourse Markers". In: E. Anne Mackay (ed.), *Orality, Literacy, Memory in the Ancient Greek and Roman World*, 35–64. Leiden and Boston 2008.

Bonifazi, Anna. *Homer's versicolored fabric: The evocative power of ancient Greek epic wordmaking.* Washington, DC and Cambridge, MA 2012.

Bonifazi, Anna, Annemieke Drummen and Mark de Kreij. Particles in Ancient Greek Discourse: Five Volumes Exploring Particle Use Across Genres. Washington, DC and Cambridge, MA. Online edition 2016. http://chs.harvard.edu/CHS/article/display/6391

Bonifazi, Anna and David F. Elmer. "Visuality in South Slavic and Homeric Epic". In: David F. Elmer and Peter McMurray (eds.), "Singers and Tales in the 21st Century: The Legacies of Milman Parry and Albert Lord". Forthcoming.

Bonifazi, Anna and David F. Elmer. "Composing lines, performing acts: clauses, discourse acts, and melodic units in a South Slavic epic song". In Elizabeth Minchin (ed.), *Orality, Literacy and Performance in the Ancient World*, 89–109. Leiden and Boston 2012a.

Bonifazi, Anna and David F. Elmer. "The Meaning of Melody: Remarks on the Performance-Based Analysis of Bosniac Epic Song". In Joseph Harris and Barbara Hillers (eds.), *Child's Children: Ballad Study and Its Legacies*, 293–309. Trier 2012b.

Chafe, Wallace L. *Discourse, Consciousness, and Time: The Flow and Displacement of Conscious Experience in Speaking and Writing*. Chicago 1994.

Collins, William. *The guritan of Radin Suane: a study of the Basemah oral epic from South Sumatra*. Leiden 1998.

Dedaić, Mirjana N. and Mirjana Mišković-Luković. *South Slavic Discourse Particles*. Amsterdam and Philadelphia 2010.

Du Bois, John W. "Argument structure. Grammar in use". In: John W. Du Bois, Lorraine E. Kumpf, and William J. Ashby (eds.), *Preferred argument structure. Grammar as architecture for function*, 11–60. Amsterdam and Philadelphia 2003.

Elmer, David F. "Presentation Formulas in South Slavic Epic Song". *Oral Tradition* 24 (2009): 41–59.

Elmer, David F. *The Poetics of Consent. Collective Decision Making & the* Iliad. Baltimore and London 2013.

Fischer, Kerstin (ed.). *Approaches to discourse particles*. Amsterdam and Boston 2006.

Foley, John Miles. *The Singer of Tales in Performance*. Bloomington 1995.

Goldman, Kenneth A. *Formulaic Analysis of Serbocroatian Oral Epic Song*. New York and London 1990.

Gumperz, John J. *Discourse Strategies*. Cambridge 1982.

Herzog, George. "The Music of Yugoslav Heroic Epic Folk Poetry". *International Folk Music Journal* 3 (1951): 62–64.

Hymes, Dell. "Particle, pause, and pattern in American Indian narrative verse". *American Indian Culture and Research Journal* 4 (1980): 7–51.

Iedema, Rick. "Multimodality, resemiotization: extending the analysis of discourse as multi-semiotic practice". *Visual Communication* 2 (2003): 29–57.

Jucker, Andreas H. and Yael Ziv. *Discourse markers: Description and Theory*. Amsterdam 1998.

Kress, Gunther R. and Theo Van Leeuwen. *Multimodal Discourse: the modes and media of contemporary communication*. London 2002.

Lenk, Uta. "Discourse markers". In: Jef Verschueren, Jan-Ola Östman, Jan Blommaert, and Chris Bulcaen (eds.), *Handbook of Pragmatics*. Amsterdam 1997.

Lord, Albert Bates *Serbocroatian Heroic Songs*. Collected by Milman Parry; edited and translated by Adam Lord. *Vol. 1*: Novi Pazar: English translations, with musical transcriptions by Béla Bartók. Cambridge and Belgrade 1954.

Maschler, Yael. *Metalanguage in interaction: Hebrew discourse markers*. Amsterdam and Philadelphia 2009.

Matras, Yaron. "Fusion and the cognitive basis for bilingual discourse markers." *International Journal of Bilingualism* 4 (2000): 505–528.

Minchin, Elizabeth. *Homer and the Resources of Memory: Some Applications of Cognitive Theory to the* Iliad *and the* Odyssey. Oxford and New York 2001.

Minchin, Elizabeth. "Spatial memory and the composition of the *Iliad*" In: E. Anne Mackay (ed.), *Orality, Literacy, Memory in the Ancient Greek and Roman World*, 9–34. Leiden and Boston 2008.

Nagy, Gregory. "Language and Meter". In: Egbert J. Bakker (ed.), *A Companion to the ancient Greek language*, 370–387. Malden, MA 2010.

Norrick, Neal. "Interjections as pragmatic markers". *Journal of Pragmatics* 41 (2009): 866–891.

Ong, Walter J. "Orality, Literacy, and Medieval Textualization". *New Literary History* 16 (1984): 1–12.

Schourup, Lawrence. “Discourse markers”. *Lingua* 107 (1999): 227–265.
Segal, Erwin M., Judith F. Duchan, and Paula Scott. “The role of interclausal connectives in narrative structuring: evidence from adults’ interpretations of simple stories”. *Discourse Processes* 14 (1991): 27–54.
Sherzer, Joel. “Poetic structuring of Kuna discourse: The Line”. *Language in Society* 11 (1982): 371–390.
Van Erp Taalman Kip, A. Maria. “*Kai mên, kai dê* and *êdê* in Tragedy and Comedy”. In: Gerry C. Wakker and Stéphanie J. Bakker (eds.), *Discourse Cohesion in Ancient Greek*, 111–133. Leiden 2009.
Viti, Carlotta. “The meanings of coordination in the early Indo-European languages”. *Revue de Sémantique et Pragmatique* 24 (2008): 35–64.
Wälchli, Bernhard. *Co-compounds and natural coordination*. Oxford 2005.
Woodbury, Anthony C. “The Functions of Rhetorical Structure: A Study of Central Alaskan Yupik Eskimo Discourse”. *Language in Society* 14 (1985): 153–190.

Mark de Kreij, Stockholm University

The priming act in Homeric epic

Abstract: Homeric epic today takes the form of a written text, but that text reflects a performance; it represents a discourse created in an exchange between performer and audience.[1] The success of this epic exchange depends on the efficient production and reception of a complex narrative composed in rich language. It is also predicated on the ability of the performer and the audience to navigate a large storyworld populated with numerous characters. This paper explores the use of fixed linguistic constructions in narrative that facilitate the cognitive process of moving from character to character and from scene to scene in Homeric epic.

The corpus of Homeric epic lends itself particularly well to a study of this phenomenon for two reasons. First, ancient Greek linguistically encodes the articulation of discourse into small steps more clearly than most western languages. Second, the Homeric corpus was composed and received in performance. The cognitive pressure involved in this process has made the discourse strongly formulaic, on a metrical, discursive, and structural level. I first present an analysis of so-called left-dislocation in Homeric Greek from a discourse perspective, with specific attention to short discourse units consisting only of a pronoun/name + particle, which I dub "priming acts." The second part of the paper analyzes two recurrent complex constructions containing priming acts that serve specific functions of (re)establishing contextual frames or accommodating a change of viewpoint. The complex constructions demonstrate the usefulness of priming acts in navigating a vast story world.

1 Discourse acts in Homeric epic

Narrative is central in human experience and interaction, since it offers an efficient way to understand, predict, evaluate, and remember complex events.[2]

---

1 For a relatively recent overview of the scholarship on the "orality" of the Homeric corpus, looking both back and forward, see issue 18.1 of *Oral Tradition* (2003), 52–86; cf. Robert Fowler, "The Homeric Question". In: *The Cambridge Companion to Homer.* Cambridge 2004, 220–232 for a concise discussion of the Homeric question with extensive literature. See for the multimodality of epic performance Bonifazi (this volume).

2 Mark Turner, *The Literary Mind.* Oxford 1996, 20. Nassim Nicholas Taleb, *The Black Swan: The Impact of the Highly Improbable.* New York 2007 illustrates how central the narrative is in

Stories, as vehicles of tradition shaped by oral transmission, may therefore provide unique insights into the workings of the mind. In his 2002 book, Foley characterizes the language of oral epic as "language, only more so."[3] The Homeric performer composed his verses on the spot, always spinning new stories with a skill born of the experience of generations of performers. The nature of the performance, a potentially endless production of verses, meant that the cognitive pressure on both performer and audience was higher than in conversation. Thus: language, only more so. We expect to see the normal cognitive processes of language production reflected in the oral narratives, only more so.[4]

To accommodate the cognitive pressure on the performer, the traditional epic is made up of themes, stock scenes with a more or less fixed sequence of events.[5] These in turn are constructed with metrical verses that accommodate small units[6] that do not necessarily coincide with syntactic units, but rather with "intonation units" (and cognates) in contemporary linguistics.[7] These intonation units are not just random subdivisions of discourse, but they actually correlate with the smallest steps in discourse; the unit was thus dubbed "discourse act" by Hannay and Kroon, among others.[8] The ancient Greek corpus is especially relevant for research into how discourse is divided in small communicative steps, because

---

human's understanding and prediction of the world, but according to Taleb this is to our own detriment.

3 John Miles Foley, *How to 'Read' an 'Oral' Poem*. Chicago, IL 2002, 127–128.

4 *Ibid.*, 128 "oral poetry is more densely coded than everyday language, in order to promote *economical exchange*." [my emphasis]

5 Albert Bates Lord, *The Singer of Tales*. Cambridge, MA 2002, 2nd edition, 68–98.

6 A large number of these acts are recurrent, repeated in the same form or different forms of the same structure, and are called formulae. Application of the term "formula" is no longer uniform, but for Homer it goes back to the work of Milman Parry, collected in Adam Parry (ed.), *The Making of Homeric Verse. The Collected Papers of Milman Parry*. Oxford 1971. Note, moreover, that formulae do not necessarily overlap with discourse acts.

7 Cf. M.A.K. Halliday and Ruqaiya Hasan, *Cohesion in English*. London 1976; Gilian Brown and George Yule, *Discourse Analysis*. Cambridge 1983; Wallace L. Chafe, *Discourse, Consciousness, and Time: The Flow and Displacement of Conscious Experience in Speaking and Writing*. Chicago, IL 1994; for Homer, see especially Egbert J. Bakker, *Poetry in Speech: Orality and Homeric Discourse*. Ithaca, NY and London 1997.

8 Cf. John McHardy Sinclair and R. Malcolm Coulthard, *Towards an Analysis of Discourse: The English Used by Teachers and Pupils*. London 1975; Eddy Roulet, "Échanges, interventions et actes de langage dans la structure de la conversation". *Études de linguistique appliquée* 44 (1981): 7–39; Caroline H. Kroon, *Discourse Particles in Latin*, Amsterdam 1995; Mike Hannay and Caroline H. Kroon, "Acts and the Relationship Between Discourse and Grammar". *Functions of Language* 12 (2005): 87–124; Kees Hengeveld and J. Lachlan Mackenzie, *Functional Discourse Grammar. A Typology-Based Theory of Language Structure*. Oxford 2008.

Greek word order roughly follows Wackernagel's law,[9] which posits that a group of words, including a range of particles,[10] tend to occur in the second position of a clause. Eduard Fränkel's adaptation of the law was that these words do not necessarily occur in the second position of a clause, but of what he called a *Kolon*, a part of the sentence that may coincide with a syntactical clause or a shorter phrase; his description of the *Kolon* can be linked quite happily to the contemporary term discourse act.[11]

The group of boundary markers identified by Wackernagel is one of the linguistic cues that I use to establish discourse act boundaries in Homeric discourse. Functionally, I define discourse acts as verbalizations of a single focus of consciousness,[12] and in form they can vary greatly in length and content.[13] I hypothesize that the emergent discourse division in the Homeric text represents

9 Jacob Wackernagel, "Über ein Gesetz der indogermanischen Wortstellung". *Indogermanische Forschungen* 1 (1892): 333–436; David Michael Goldstein, *Wackernagel's Law in Fifth-Century Greek*, doctoral thesis, UC Berkeley 2010, 9 offers a recent discussion.

10 ἄρα (ára), γάρ (gár), δέ (dé), δή (dḗ), μάν (mán), μέν (mén), μήν (mḗn), οὖν (oûn), ὦν (ôn).

11 Eduard Fraenkel, "Kolon und Satz II. Beobachtungen zur Gliederung des antiken Satzes". In: *Nachrichten der Göttinger Gesellschaft der Wissenschaften, Phil.-hist. Klasse*, 1933, 319–354. The term *Kolon*, or colon, was originally (and still often is) used to denote metrical units, which is reflected in the idea of colometry, for example. See Frank Scheppers, *The Colon Hypothesis. Word order, discourse segmentation and discourse coherence in Ancient Greek*. Brussels 2011 for the most recent work on the colon as a subdivision of Greek prose, and e.g. Bruno Gentili and Liana Lomiento, *Metrica e ritmica: storia delle forme poetiche nella Grecia antica*. Milan 2003 for more on the metrical colon. Recent works that draw on Fraenkel are Helma Dik, *Word Order in Ancient Greek: a Pragmatic Account of Word Order Variation in Herodotus*. Amsterdam 1995 and *Word Order in Greek Tragic Dialogue*. Oxford 2007; Egbert J. Bakker, *Linguistics and Formulas in Homer: scalarity and the description of the particle "per"*. Amsterdam 1988 and *Poetry in Speech*; Goldstein, *Wackernagel's Law*; Nicolas Bertrand, *L'ordre des mots chez Homère: Structure informationnelle, localisation et progression du récit*. Unpublished thesis, Université de Paris-Sorbonne 2010; Scheppers, *The Colon Hypothesis*.

12 I take the definition from Chafe, *Discourse, Consciousness, and Time*, 63, who applies it to intonation units: "...an intonation unit verbalizes the speaker's focus of consciousness at that moment."

13 This flexibility in form and content is parallelled fully in the intonation units Chafe finds in his *Pear Stories*: Wallace L. Chafe, *The Pear Stories: Cognitive, Cultural, and Linguistic Aspects of Narrative Production*. Norwood, NJ 1980. For Homer, I posit that discourse acts are in principle never longer than one hexameter verse, and generally shorter. Verse end commonly coincides with discourse act boundary, and in those cases where it does not, the discourse act that carries over to the next verse never reaches as far back as verse beginning. Even in those cases, however, verse end may still have been noticeable; cf. Frederick Turner and Ernst Pöppel, "The Neural Lyre: Poetic Meter, the Brain, and Time". *Poetry* 142 (1983): 286 "[t]he LINE is preceded and followed by a distinct pause (...), which (...) divides the verse into clearly identifiable pieces" [capitals original].

the reality of performance: discourse act boundaries will have been accompanied by a pause or a discontinuity of intonation contour.[14]

2 Priming acts

Among the many forms that discourse acts can take, there is one kind that is often ignored or regarded as problematic. Consider the following example from the *Iliad*:

(1) *Iliad* 15.390–392
Πάτροκλος δ' |[15] εἷος μὲν Ἀχαιοί τε Τρῶές τε
τείχεος ἀμφεμάχοντο | θοάων ἔκτοθι νηῶν, |
τόφρ' ὅ γ' | ἐνὶ κλισίῃ ἀγαπήνορος Εὐρυπύλοιο |

And Patroclus, | as long as the Achaeans and Trojans
were fighting around the wall, | far from the ships, |
all the while HE, | in the tent of pleasant Eurypylus, |

In line 390, we find a referent, activated with a name: Patroclus, followed by the particle δέ (dé). After this solitary name in the nominative follows a new discourse act, as is marked by the particle μέν (mén), which is restricted to second position in the act. Syntactically, the nominative Patroclus is not the subject of a main verb, since in line 392 a new nominative is introduced: ὅ (hó), he, which will become the subject of the main clause to follow in line 393. With few exceptions[16] independent nominatives like this have been called "pendant nominatives,"

---

14 Bakker, *Poetry in Speech*, 101 argues that second-position words (also called postpositives) mark intonational boundaries, but he does not apply this criterium consistently (see the examples on pages 97 and 110). See Andrew M. Devine and Laurence D. Stephens, *The Prosody of Greek Speech*. New York, NY 1994 for an attempt at establishing how ancient Greek was rendered prosodically, and especially chapter 9 (409–455) for a discussion of what they call major-phrase boundaries.

15 The bars here and in all following examples mark discourse act boundaries, established because of the presence of boundary markers (as μέν here) or because of the independence of a phrase or clause (θοάων ἔκτοθι νηῶν and ἐνι κλισίῃ ἀγαπήνορος Εὐρυπύλοιο are adverbial phrases, for example).

16 See Simon R. Slings, "Written and Spoken Language: an exercise in the pragmatics of the Greek sentence". *Classical Philology* 87 (1992): 95–109 on information status; Bakker, *Poetry in Speech*, calls them "frames"; Bertrand, *L'ordre des mots*, proposes a pragmatic (in the narrow sense of Functional Grammar) approach.

while the (pro)noun is called "proleptic"[17] if the referent is in another case than the nominative. This view has persisted up to present day, as illustrated by De Jong's comment on a similar instance: "[οἱ δ' is] a pendant nominative (or frame) (...). This is a fairly common phenomenon in the Homeric epics (...), a clear manifestation of their oral syntax."[18]

Constructions of this nature would be treated under the header "left-dislocation" in contemporary linguistics.[19] In such constructions, the subject of a proposition is presented as a separate intonation unit, followed by a proposition with the subject repeated or changed. Chafe observed the same phenomenon in his analysis of oral narrative: "It is not unusual for an intonation unit to verbalize little or nothing more than a referent," and he calls the resulting short intonation units "isolated referents."[20]

---

17 E.g. Pierre Chantraine, *Grammaire Homérique II: Syntaxe*. Paris 1963; Cornelis J. Ruijgh "La place des enclitiques dans l'ordre des mots chez Homère d'après la loi de Wackernagel". In: Heiner Eichner and Helmut Rix (eds.), *Sprachwissenschaft und Philologie: Jacob Wackernagel und die Indogermanistik heute*. Wiesbaden 1990, 213–233; Irene J.F. de Jong, *Homerus:* Iliad *book XXII*. Cambridge 2012.

18 De Jong, *Iliad* book XXII, 122 on *Iliad* 22.248–9: οἱ δ' ὅτε δὴ σχεδὸν ἦσαν ἐπ' ἀλλήλοισιν ἰόντες / τὸν πρότερος προσέειπε μέγας κορυθαίολος Ἕκτωρ. Compare Ivo Hajnal, "e-ke-qe oder e-ke?– Mündlichkeit und Schriftlichkeit in den mykenischen Palastarchiven". In: Thomas Krisch, Thomas Lindner, and Ulrich Müller (eds.), *Analecta Homini Universalis Dicata. Arbeiten zur Indogermanistik, Linguistik, Philologie, Politik, Musik und Dichtung. Festschrift für Oswald Panagl zum 65. Geburtstag*. Stuttgart 2004, 245, who finds in similar constructions in Linear B "...Spuren mündlicher Syntax."

19 The term goes back to Henk van Riemsdijk and Frans Zwarts, "Left dislocation in Dutch and the status of copying rules". In: Elena Anagnostoupou et al. (eds.), *Materials on Left Dislocation*. Amsterdam 1997 [1974], 13–31; more recent work in Ronald Geluykens, *From discourse process to grammatical construction: on left-dislocation in English*. Amsterdam 1992; and William Salmon, "Double Subjects and Verbal Demonstrations". *Journal of Pragmatics* 42 (2010): 3431–3443; for a critique of the term see Elinor Ochs Keenan and Bambi Schieffelin, "A Reconsideration of Left Dislocation in Discourse". In: Proceedings of the 2nd Annual Meeting of the Berkeley Linguistics Society 1976, 240–257 and more recently Simona Pekarek Doehler, "Emergent Grammar for All Practical Purposes: The on-Line Formating of Dislocated Constructions in French Conversation". In: Stefan Pfänder and Peter Auer (eds.), *Constructions: Emerging and Emergent*. Berlin and New York 2011, 46–88, with extensive bibliography.

20 Chafe, *Discourse, Consciousness, and Time*, 67, and on 68: "Typically, such *isolated referents* (expressed as so-called free NPs) are subsequently included as participants in events and states. But intonation units like these show that it is quite possible for speakers to focus on a referent alone." Chafe (*ibid.*, 108–119) also argues that each intonation unit can bring no more than one new idea (the one-new-idea constraint), so if there is a lot of cognitive processing involved in retrieving a referent, it makes sense that an intonation unit consists of no more than an isolated referent.

Chafe's work forms an important source of inspiration for Bakker's reading of Homer as a string of short intonation units. Bakker calls Πάτροκλος δέ in (1) a "framing unit," which establishes the referents of the noun or pronoun as a "frame" for the upcoming discourse, "the demarcation of a frame limiting one's field of vision for the next moments or speech units, the area within which addition of detail can meaningfully take place." This construction is a manifestation of "the general movement of Homeric discourse (...) from the global to the specific, and from the framing to the framed."[21]

What all these treatments of short referential discourse acts or intonation units have in common is that they describe them as forward-looking. It is a pendant nominative "waiting" for its verb, an isolated referent who will be in focus in the following, a unit dislocated to the left of (i.e. before) its host construction, or units "serving as frames for what lies ahead."[22] This tendency would appear to be a result of our bird's-eye view of text, looking at discourse as a finished construct; we can identify different components and decide what lies in the center and what lies in the periphery of these components. However, in practice we receive discourses linearly, reading from left to right, or hearing it from beginning to end as in the case of the Homeric epics. When we are presented with a discourse act, it will add to our knowledge, trigger inferences about the upcoming discourse, but it will also have an effect *in situ*, and it is the latter element that I focus on. Irrespective of what follows, the cognitive processing involved in this narrative step justifies the devotion of an entire discourse act to no more than naming a character.[23] Such acts prime a character, and because of how the participants in a narrative are organized in our minds they also prime the time and place that the character is associated with—in the case of (1) the Greek camp—and it is because of this double potential that I call them priming acts. Semantically, the constituents do nothing more than retrieve or establish a referent, but the construction reveals some of the efficiency of the human mind in keeping track of a long narrative taking place in a complex storyworld.

---

21 Bakker, *Poetry in Speech*, 86–111, both quotes from page 89. Note that De Jong refers to Bakker's sense of frame in her commentary (De Jong, *Iliad* book XXII, 122).

22 Bakker, *Poetry in Speech*, 110. Although it comes from a different approach, the term "anticipation" used by Matthew Clark, *Out of Line. Homeric Composition Beyond the Hexameter*. Lanham, MD 1997 for constructions like αὐτὰρ Ἀθήνη at verse end might be added to this list, and compare Mark W. Edwards, *Sound, Sense, and Rhythm: Listening to Greek and Latin Poetry*. Princeton, NJ 2002, 44 on αὐτὰρ Ἀχιλλεύς: "...our attention is directed toward (i.e. the poet zooms in on) a particular individual upon whom will center the action to come."

23 About similar instances, Bakker says (*Poetry in Speech*, 110): "In both cases a separate unit is devoted to the narrative act of returning to Odysseus and Achilles, reflecting the cognitive effort that is involved in the reactivation."

3 Moving the mind's eye

In the quote above, Bakker describes the function of his "frames" in terms of the "field of vision." The manipulation of viewpoint in narrative discourse is in fact an important instrument in the movement through the narrative space, a topic that has attracted considerable attention in recent decades.[24] I understand viewpoint in the sense of Talmy, as a real viewpoint experienced by the mind's eye:

> [I]n a narrative, we can apply all our usual conceptions of spatial relations to understanding the location and motion of a viewpoint, the angle and direction of our sighting from that viewpoint to an entity to which we are attending, and the size and shape of this attentional projection.[25]

In his 2002 *Story Logic*, Herman says: "framing representations of the where in a story is a major dimension of narrative processing. At stake is much more than simply filtering out descriptive detail to form interpretations of 'core' narrative elements."[26]

Priming acts form part of these "framing representations of the where in the story," guiding speaker and hearer through abstract, often relative spaces in the storyworld. This framing is established and navigated much more through linguistic form than through propositional content, which correlates with the fact that explicit reference to space is so rare in Homeric oral epic.[27] For a narrative

---

24 David Herman, *Story Logic. Problems and possibilities of narrative.* Lincoln 2002, 263–285 gives an overview of literature on space in narratology. For Homer, the most recent works are Alex C. Purves, *Space and Time in Ancient Greek Narrative.* Cambridge 2010; Jenny Strauss Clay, *Homer's Trojan Theater: Space, vision, and memory in the* Iliad. Cambridge 2011; Christos Tsagalis *From Listeners to Viewers: Space in the* Iliad. Washington, DC 2011; Irene J.F. de Jong (ed.), *Space in Ancient Greek Literature.* Leiden 2012.

25 Leonard Talmy, "A Cognitive Framework for Narrative Structure". In: Leornard Talmy, *Toward a Cognitive Semantics 2.* Cambridge, MA 2000, 437.

26 Herman, *Story Logic*, 270; note that *Space in Ancient Greek Literature* volume references neither Tsagalis, *From Listeners to Viewers*, nor Barbara Dancygier, *The Language of Stories: A Cognitive Approach.* Cambridge 2012, Catherine Emmott, *Narrative Comprehension. A Discourse Perspective.* Oxford 1997, or Herman, *Story Logic*, in the bibliography, yet in the introduction De Jong (1) speaks of "the relative neglect of space in narratological theory."

27 De Jong (ed.), *Space in Ancient Greek Literature*, 22, with note 1; see also Theodore Murdock Andersson, *Early Epic Scenery: Homer, Virgil, and the Medieval Legacy.* Ithaca, NY 1976; Irene J.F. de Jong, René Nünlist, and Angus M. Bowie (eds.), *Narrators, Narratees, and Narratives in Ancient Greek Literature.* Leiden 2004; but see Clay, *Homer's Trojan Theatre*, who shows convincingly that the (absolute) geography of the *Iliad* makes sense, despite claims to the contrary by other critics. The relevance of this geography to performer and audience is doubtful, however, see Marie-Laure Ryan, "Cognitive Maps and the Construction of Narrative Space". In: David Herman (ed.),

action to take place we must presuppose a place, and for a longer narrative like the *Iliad* or *Odyssey* to happen we must presuppose a world, a storyworld, and the strength of this presupposition allows the author or performer to construct, adapt, and refer to it at will. The storyworld is encoded explicitly only when the narrator or performer needs it,[28] but the (implicit) spatial dimension of the narrative constantly determines its linguistic realization. It influences lexical choices (e.g. "coming" versus "going") and syntax (e.g. active versus passive constructions), but it is also reflected in discourse progression. Priming acts, short discourse acts devoted to the activation of a referent, accommodate the journey through the visual representation of the narrative, establishing a path through the storyworld.[29] In terms of viewpoint, priming acts provide focal points around which the mind's eye navigates the epic world,[30] moving away from characters or toward them, and establishing them as anchors for their surroundings. One may

---

Narrative Theory and the Cognitive Sciences. Stanford, CA 2003, 283 "...people read for the plot and not for the map, unless they are literary cartographers. We construct mental models of narrative space *only as far as we find a cognitive advantage in this activity*—only as far as is needed to achieve immersion in the textual world." [my emphasis]

28 This is the approach taken in De Jong (ed.), *Space in Ancient Greek Literature*: all contributions consider references to space or items-in-space first and foremost as narrative functions.

29 This reflects the hodological (as opposed to cartographic) view of narrative, on which see e.g. Gabriel Zoran, "Towards a Theory of Space in Narrative". *Poetics Today* 5.2 (1984): 309–335; Barbara Landau and Ray Jackendoff, "'What' and 'Where' in Spatial Language and Cognition". *Behavioral and Brain Sciences* 16 (1993): 217–265, Herman, *Story Logic*, and Ryan, "Cognitive Maps"; applied to classics by Anna Bonifazi, "Memory and Visualization in Homeric Discourse Markers". In: E. Anne Mackay (ed.), *Orality, Literacy, Memory in the Ancient Greek and Roman World*. Leiden 2008, 35–64; Anna Bonifazi, *Homer's Versicolored Fabric. The Evocative Power of Ancient Greek Epic Word-Making*. Washington, DC 2012; Elizabeth Minchin, "Spatial Memory and the Composition of the *Iliad*". In: Anne Mackay (ed.), *Orality, Literacy, Memory in the Ancient Greek and Roman World*. Leiden 2008, 9–34; Purves, *Space and Time*; Clay, *Homer's Trojan Theatre*. Consider also Christopher Collins, *The Poetics of the Mind's Eye: Literature and the Psychology of Imagination*. Philadelphia, PA 1991, 98 "the consecutiveness of speech accords with the consecutiveness of visual perception."

30 For narrative as a visual experience, cf. Monika Fludernik, *Towards a 'Natural' Narratology*. London 1996; Emmott, *Narrative Comprehension*; Herman, *Story Logic*; Barbara Dancygier and Eve Sweetser (eds.), *Viewpoint in Language*. Cambridge 2012. For Homeric narrative, see Bakker, *Poetry in* Speech, 43; Eric A. Havelock, *Preface to Plato*. Cambridge, MA 1963, 188 "[epic] units of meaning are highly visualized in order that vision may lead on to vision"; David C. Rubin, *Memory in Oral Traditions: The Cognitive Psychology of Epic, Ballads, and Counting-Out Rhymes*. New York, NY 1995, 59 "oral traditions appear to be remarkably spatial"; Minchin , "Spatial Memory", 10 "the narrative of oral epic has a very real spatial dimension"; Clay, *Homer's Trojan Theatre*, 110 "the poet sees his story, and his narrative translates what he sees into words that in turn evoke a vision of the events for his audience."

compare Landau and Jackendoff's description of space in narrative as regions occupied by landmarks (reference objects).[31]

4 The Homeric priming act

The most versatile Homeric priming acts take the form of noun phrase in the nominative + δέ (see example 1) or αὐτὰρ/ἀτὰρ + NP in the nominative.[32] The latter generally occurs at verse end, not infrequently just before direct speech,[33] but there are also cases where a boundary marker in the following discourse act shows intonational division, as in (2):[34]

(2) *Odyssey* 17.64–67
τὸν δ' ἄρα πάντες λαοὶ ἐπερχόμενον θηεῦντο. |
ἀμφὶ δέ μιν μνηστῆρες ἀγήνορες ἠγερέθοντο |
ἔσθλ' ἀγορεύοντες, | κακὰ δὲ φρεσὶ βυσσοδόμευον. |
αὐτὰρ ὁ | τῶν μὲν ἔπειτα ἀλεύατο πουλὺν ὅμιλον, |

And him [Telemachus], then, all the men admired as he came in. |
And around him the proud suitors pressed together |
saying good things | and pondering bad things in their minds. |
But he | he then avoided their great multitude |

In (2) Telemachus is the one referred to by the combination of ἀυτάρ + pronoun, even though he must be regarded as completely accessible in this passage. The narrator has just described how Penelope's suitors fuss about him, so there is

31 Landau and Jackendoff, "'What' and 'Where'", 223.
32 I have limited my analysis of the construction to NPs in the nominative that refer to a character. The examples of priming acts with αὐτάρ/ἀτάρ followed by a boundary marker or independent construction are: *Iliad* 5.29, 5.327, 5.849, 7.383, 10.488, 11.732, 15.630, 16.220, 19.15, 20.423, 20.484, 21.71, 21.205, 21.550, 23.1, 23.110, 23.798, 24.244; *Odyssey* 1.57, 6.2, 8.83, 9.431, 10.261, 14.42, 14.245, 14.480, 16.454, 17.67, 17.360, 18.66, 18.69, 19.209, 21.404, 22.116, 24.309, 24.367, 24.392. αὐτάρ followed only by verse end: *Iliad* 1.127, 1.348, 1.430, 5.308, 5.844, 8.268, 9.628, 10.154, 11.321, 11.326, 11.341, 11.489, 11.664, 11.759, 11.762, 12.78, 13.164, 15.405, 16.124, 16.728, 17.266, 18.148, 18.231, 18.314, 19.83, 19.188, 19.268, 19.351, 20.75, 20.283, 20.441, 21.520, 21.538, 22.3, 23.128, 23.257, 24.3; *Odyssey* 2.182, 5.370, 7.81, 8.40, 8.367, 8.521, 13.28, 13.367, 14.30, 14.171, 16.119, 16.177, 18.100, 18.394, 19.388, 19.479, 22.292, 22.493, 23.67, 23.181, 24.220. ἀτάρ followed only by verse end: *Iliad* 3.268, 11.153, 11.744, 17.580, 18.223; *Odyssey* 18.99.
33 Priming acts regularly occur near the beginning of direct speech, which is no doubt connected to the importance of establishing the source of direct speech in an epic performed by a single voice.
34 See Bakker, *Poetry in Speech*, 101.

little cognitive action involved in focusing on Telemachus. Still, the priming act must have a function beyond simply establishing the subject of the upcoming sentence, since the switch of subject could have been accomplished with simply a third-person singular verb form. Here, the priming act is not used to redirect attention to a new referent, but to change the viewpoint within the scene. The combination of the discourse marker αὐτάρ plus the nominative (after three accusatives: τόν, ἐρχόμενον, μιν) shifts the viewpoint from outside, regarding the scene from an external viewpoint, to looking over Telemachus' shoulder, or even through his eyes, as he ignores the suitors.[35] Regardless of its exact form, the priming act always has as its function the reorientation of the cognitive viewpoint, a movement of the mind's eye, be it toward a new character and/or narrative space as in (1), or a change of viewpoint within a narrative space as in (2).[36]

Priming acts are a verbally economical way of redirecting attention, since they transport the narrative to a new place with only the mention of a character rather than "Meanwhile on the Olympus..." In her 1997 book, Emmott has done crucial work to illuminate the cognitive workings of such narrative transitions. She speaks of a narrative world containing "contextual frames,"[37] places in the storyworld that contain certain characters, who are "bound" to that place until they leave it.[38] As long as the current narrative takes place within that frame (A), the referents "bound" to it remain active. When the performer activates another contextual frame (B), the characters bound to frame (A) become inaccessible, but they remain bound to frame (A), and are immediately accessible if frame (A) is reactivated. For example, when in book 16 of the *Iliad* the Homeric performer tells of the battle between Sarpedon and Patroclus, then shifts to the Olympus where

---

35 Bonifazi, *Homer's Versicolored Fabric*, 218–243 argues for αὐτάρ that rather than a marker of adversativity on a propositional level (to be translated as "but") it seems to work on the level of the presentation of narrative, adjusting the viewpoint by switching to a new referent or zooming in on one of a group. On page 222 she speaks of "the visual and presentational function of αὐτάρ."

36 For the function of αὐτάρ in viewpoint switches, consider also *Odyssey* 17.254–255: ὣς εἰπὼν τοὺς μὲν λίπεν αὐτόθι ἦκα κιόντας, / αὐτὰρ ὁ βῆ... *he left them, αὐτάρ he went.*

37 Emmott's contextual frames are an application of the idea of [semantic] frames—interconnected semantic networks, on which see Charles J. Fillmore, "The need for a frame semantics within linguistics". In: Hans Karlgren (ed.), *Statistical Methods in Linguistics* 12 (1976): 5–29—to characters in narrative discourse.

38 Emmott, *Narrative Comprehension*, 121–122. Emmott's point can be linked directly to the observation by the psychologists Eugene Winograd and Vaughan E. Church, "Role of Spatial Location in Learning Face-Name Associations". *Memory and Cognition* 16.1 (1988): 6–7 (referred to by Minchin, *Spatial Memory*, 10), who claim that spatial information can cue the recall of associated material.

Zeus decides the fate of his son Sarpedon, the frame containing Patroclus and Sarpedon fighting remains intact, and available for reference after the scene on the Olympus. In fact, the transition back to the battlefield is accomplished by a priming act (16.642 οἱ δ' | And they [sc. Sarpedon and Patroclus], see example 4 below). Once characters are bound to a location, activating that character gives both performer and audience access to the entire contextual frame.

5 Analysis I: The character as a guide

The epic heroes are central in Homeric narrative, and their movements or perceptions guide us through the storyworld.[39] Even if it is not the physical movement of a character that leads the audience to a new scene or viewpoint, the act of activating them may be just as effective. This makes priming acts versatile tools in the navigation of the storyworld, which led to their inclusion in a number of specialized constructions in the Homeric epics. Homeric language is strongly formulaic, full of constantly recurring semantic and structural packages.[40] Little groups of words have come to function as hyperlinks to whole bodies of knowledge, so that an epithet of a hero can invoke a whole narrative on its own.[41] A closer inspection of priming acts yields another kind of formulaicness, where a recurrent group of words does not invoke a body of meaning, but serves to accommodate a specific cognitive action. Not only the content of the epic storyworld but even the navigation through it has been formalized in Homeric language.[42]

---

39 Scott Richardson, *The Homeric Narrator.* Nashville, TN 1990, 110 "...the Homeric narrator, though he recognizes his privilege of unfettered movement, usually declines it and lets himself be led from scene to scene by one of his characters." See on this topic also Brigitte Hellwig, *Raum und Zeit im homerischen Epos.* Tübingen 1964, 88–115 and Edwards, *Sound, Sense, and Rhythm*, 52: "Smoothly and uninterruptedly, action follows upon action, with continuity, not division, being always the poet's aim."
40 On a larger scale epic is "formulaic" too, consisting of recurring themes (Lord, *Singer of Tales*) and type scenes (Walter Arend, *Die typische Szenen bei Homer.* Berlin 1933). Elizabeth Minchin, *Homer and the Resources of Memory: Some Applications of Cognitive Theory to the* Iliad *and the* Odyssey. Oxford 2001 shows convincingly how "themes" in oral poetry might be analyzed with the terminology of scripts, which is set out in Roger C. Schank and Robert P. Abelson, *Scripts, Plans, Goals, and Understanding: An Inquiry into Human Knowledge Structures.* Hillsdale, NJ 1977.
41 See on such signs in Homeric language especially John M. Foley, *Homer's Traditional Art.* University Park, PA 1999, 26–31.
42 Bonifazi [this volume] captures this aspect of epic language well, when she says: "Metanarration is as traditional as narration," that is, not only are the stories encoded in the traditional language, even the *way of telling* the stories is encoded in the traditional language.

In the following example, the priming act is incorporated in a larger construction that includes what is formally a temporal subclause. After hearing about the situation at the Greek camp, where Achilles is seething because Agamemnon has wronged him, the audience is brought to Odysseus on his way by ship to appease Chryses, priest of Apollo, by returning his daughter:

(3) *Iliad* 1.430–433

| αὐτὰρ Ὀδυσσεὺς |
ἐς Χρύσην ἵκανεν | ἄγων ἱερὴν ἑκατόμβην. |
οἳ δ' | ὅτε δὴ λιμένος πολυβενθέος ἐντὸς ἵκοντο |
ἱστία μὲν στείλαντο, | θέσαν δ' ἐν νηῒ μελαίνῃ, |

| And Odysseus |
arrived at Chryses | bringing the sacred hecatomb. |
And they | when they arrived within the deep harbour |
they furled the sails | and put them away in the black ship |

The transition to a new location is accommodated by the priming act αὐτὰρ (autár) Ὀδυσσεύς, followed by a line which functions as a kind of abstract to remind the hearer of what Odysseus was supposed to be doing. Then follows the priming act οἳ δέ (hoì dé), as part of a longer verse that captures the moment of arrival at the place for the upcoming action. The pronoun οἵ is a nominative plural, which is instructive in two ways. First, it shows that the mention of only Odysseus has activated a whole group which can then be referred to with nothing more than a pronoun. Second, it explains why the priming act is necessary: here the pronoun does not retrieve a referent so much as create it. Odysseus and his unnamed crew are set up as a pool of accessible referents, and the verse as a whole serves to place Odysseus, his men, and Chryses' daughter on the shore for the crucial event of appeasing the priest of Apollo by returning his daughter and making an offering to the god.

In (3) the pronoun refers to an undefined collective, but more frequently it demarcates a specific number of referents, like in (4). Zeus has just decided the fate of his son Sarpedon who is fighting Patroclus, and in line 462 the narrative returns to the battlefield:[43]

43 As often, the movement to a new place in the storyworld is not completely independent. Along with the bloody rain, the audience moves from above (the Olympus) down to the battlefield. This is an inverse parallel of the tendency in Homer to accompany a transition from the battlefield to the world of the gods with an act of visual or aural perception (see note 58 below for some examples).

(4) *Iliad* 16.459–463
αἱματοέσσας δὲ ψιάδας | κατέχευεν ἔραζε |
παῖδα φίλον τιμῶν, | τόν οἱ Πάτροκλος ἔμελλε
φθίσειν | ἐν Τροίῃ ἐριβώλακι | τηλόθι πάτρης. |
οἳ δ' | ὅτε δὴ σχεδὸν ἦσαν | ἐπ' ἀλλήλοισιν ἰόντες, |
ἔνθ' ἤτοι Πάτροκλος

And bloody rain | he [Zeus] poured on the earth |
honouring his beloved son | whom Patroclus would
kill | in deep-soiled Troy | far from his fatherland. |
And they | when they were close | bearing down on each other |
right then Patroclus

The verse in line 462 above recurs eleven times (all in the *Iliad*) and De Jong notes rightly that it "typically introduces martial duels."[44] Here it also marks the transition between two spaces in the storyworld: the Olympus where Hera and Zeus converse, and the battlefield in front of Troy. The performer smooths the transition by first mentioning Sarpedon and Patroclus as the objects of Zeus' thoughts and then priming them as actants with οἳ δέ (hoì dé).

The function of this recurrent verse is to establish the two referents as the main actants in the upcoming narrative, but also to create a place of action. Being on the battlefield, it is hard to imagine an absolute location, so this verse serves to provide a temporary narrative space.[45] It sets apart a section of the battlefield for the fighters to finish their duel before the victor moves on. This temporary demarcated space functions as a contextual frame, set up by the verse-long construction.[46] The cue for the cognitive act of binding characters to this location

44 De Jong, *Iliad book XXII*, p. 122. She comes to a total of 12 instances, but there are only 11: *Iliad* 3.15, 5.14, 5.630, 5.850, 6.121, 11.232, 13.604, 16.462, 20.176, 21.148, 22.248. The one other she adduces (23.816) is in an athletic contest, as she notes, but has the form ἀλλ' ὅτε δή κτλ., i.e. without the priming act οἳ δέ.

45 *Iliad* 21.148 is an exception, in that it locates the action clearly on the banks of the river Xanthus.

46 Besides the recurrent verse (see note 44), compare these other instances where the referents are bound to a temporary location: *Iliad* 4.446 οἵ δ' | ὅτε δή ῥ' ἐς χῶρον ἕνα ξυνιόντες ἵκοντο; *Iliad* 10.526 οἳ δ' | ὅτε δή ῥ' ἵκανον | ὅθι σκοπὸν Ἕκτορος ἔκταν; *Iliad* 18.520 οἳ δ' | ὅτε δή ῥ' ἵκανον | ὅθι σφίσιν εἶκε λοχῆσαι. When the space is a geographic location, like in (3), the binding of characters to a certain place is even more explicit: *Iliad* 7.113 οἳ δ' | ὅτε δὴ κλισίῃσιν ἐν Ἀτρεΐδαο γένοντο; 10.180 οἵ δ' | ὅτε δή φυλάκεσσιν ἐν ἀγρομένοισιν ἔμιχθεν; *Iliad* 11.618 οἳ δ' | ὅτε δὴ κλισίην Νηληϊάδεω ἀφίκοντο; *Iliad* 23.38 οἳ δ' | ὅτε δὴ κλισίην Ἀγαμέμνονος ἶξον ἰόντες; *Odyssey* 1.126 οἱ δ' | ὅτε δή ῥ' ἔντοσθεν ἔσαν δόμου ὑψηλοῖο; *Odyssey* 16.324 οἱ δ' | ὅτε δὴ λιμένος πολυβενθέος ἐντὸς ἵκοντο; *Odyssey* 24.362 οἱ δ' | ὅτε δή ῥ' ἵκοντο δόμους ἐῢ ναιετάοντας. Finally, there are five instances in the Odyssey where Penelope reaches a new place, binding her into that contextual frame: *Odyssey* 1.332, 7.3, 18.208, 21.42, 21.63.

is also encoded linguistically, through the consistent use of a verb of arriving or coming to a place.[47] By combining priming acts with a specific kind of verb construction, this set of verses encode the establishment of a contextual frame.

In terms of viewpoint, the effect of the construction is perhaps best imagined as a slight zooming out of the mind's eye from the individuals to create something of a boxing ring with clear boundaries. The effect created is one of slowing down and speeding up of time: as long as the duel lasts, the battlefield around it is completely out of focus and time seems to stand still. The transgression of the boundaries, often after the death of one of the combatants, will signal the dissolution of the contextual frame. Before the frame is dissolved, there may also be a final conversation between the victor and his victim, after which action is resumed. At this moment the victor leaves, or another character intrudes, such as the charioteer or the horses of the fallen hero. The contextual frame is discrete as long as it is needed, with a suspension of everything around it assumed, until the battlefield is allowed to intrude into the frame and time "speeds up" again.[48]

6 Analysis II: Following the hero's gaze

The role of priming acts in reorienting viewpoint emerges even more clearly from one final set of examples: transitions based on characters' perceptions.[49] Richardson remarks that Homer prefers natural transitions, letting a character's movements or perception guide the journey through the storyworld. The number of transitions implicitly or explicitly effected through sensory perception is very large, so it is not surprising that this very effect has been fixed in a verse-long construction.[50]

---

47 The semantics of "coming" and "going" can be encoded in Greek verbs, but they need not be. In the examples in footnote 46 above the verb used is a form of ἱκνέομαι, *to arrive* in 7 out of 10 cases. In all other instances and in the 11 instances of the recurrent verse (note 44) the sense is one of characters coming nearer to one another (as marked by σχέδον, *near* or ἐν/ἔντοσθεν, *inside*) in a specific place.

48 One may compare the effect created at the moment of a successful assassination in the videogame *Assassin's Creed*: the camera zooms in on the assassin and his victim, the surroundings are blurred, and the two protagonists have a final conversation.

49 Compare Bonifazi's "camera-in-the-mind" [this volume].

50 Richardson, *The Homeric Narrator*, 100. Cf. Irene J. F. de Jong, *Narrators and Focalizers. The presentation of the story in the* Iliad. London 2004, 2nd edition, 105–107, where she argues that perceptions (105) "form the hinges of the story" and can be used (107) "to bring on stage

At the beginning of book 3 of the *Iliad* the Greek and Trojan armies meet for the first time on the battlefield, setting up their troops across from each other. The narrator then tells us that Paris (Alexander) led the Trojans, and came forth from the army. While Paris is still in focus, the shift to the Greek army is established by making Paris the object of someone else's gaze:

(5) *Iliad* 3.21–22
τὸν δ' | ὡς οὖν ἐνόησεν ἀρηΐφιλος Μενέλαος |
ἐρχόμενον προπάροιθεν ὁμίλου | μακρὰ βιβάντα, |

And him [sc. Alexander], | just as Menelaus, beloved to Ares, saw him |
coming in front of the crowd, | taking long strides, |

This act of perception leads us to the Greek army, where Menelaus is eager to do battle with Paris:

(6) *Iliad* 3.29–31
αὐτίκα δ' ἐξ ὀχέων | σὺν τεύχεσιν ἆλτο χαμᾶζε. |
τὸν δ' | ὡς οὖν ἐνόησεν Ἀλέξανδρος θεοειδὴς |
ἐν προμάχοισι φανέντα, | κατεπλήγη φίλον ἦτορ, |

and at once from the car | in his armour he [sc. Menelaus] leapt to the ground.
And him, | just as god-like Alexander saw him, |
appearing among the vanguard, | he [sc. Alexander] was scared in his heart, |

Menelaus vaults from his chariot eager to engage the man who stole his wife, and pushes through the throng to the front of the army. The priming act τὸν δέ (tòn dé) in the accusative moves the viewpoint away from Menelaus and makes him in turn the object of Paris' gaze. Thus, the verse brings the audience back to the Trojans, where Paris retreats into the mass of Trojan soldiers, later to be scolded by his brother Hector (line 38).

This visual back-and-forth is a dynamic way of establishing the two antagonists facing off on the battlefield, both leaders at the same time representing their respective armies, but it also serves a function in the larger narrative. The reason for the Trojan War was the taking of Menelaus' wife Helen by Paris. In narrated time, we must assume that the two have met before (since the *Iliad* only covers the final weeks of a ten-year conflict), but in the plot of the *Iliad* this is the first moment that the two characters are in the same narrative space; it is in fact Paris'

characters necessary for the story." See also Hedwig Jordan, *Der Erzählungsstil in den Kampfszenen der* Ilias. Breslau 1905 and Hellwig, *Raum und Zeit*.

first appearance. The moment that they see each other is thus drawn out in two constructions, where first Menelaus sees Paris and then Paris sees Menelaus. Their subsequent different reactions to seeing the other, extended by similes, are crucial ethopoetic moments for these characters in the *Iliad*.[51]

The form of this verse-long construction is always that of a (pro)noun + δ' | ὡς οὖν (d' hōs oûn) + a verb of perception + the subject or object, and it recurs seventeen times in the narrator text of the *Iliad* and seven times in the *Odyssey*.[52] Depending on the case of the (pro)noun, the mind's eye moves away from (oblique case) or towards (nominative) the referent in the priming act. It can mark a change of viewpoint within one contextual frame, or, as here, effect the transition from one contextual frame to another, when a character in one frame perceives one in another.[53]

7 The priming act as a narrative tool

The priming act is indispensable when dealing with a storyworld conceived and imagined in performance. In Homeric epic it is a verbally economical[54] tool for the performer to manipulate the mind's eye of his audience in the storyworld, by directing attention to new places or offering a new vantage point within the same scene. The use of priming acts exploits the mind's capability of binding characters to places in the same way that whole semantic fields are bound together in semantic frames.[55] It emerges from the material that priming acts are inherent parts of a prevalent technique in the Homeric narrative to move the mind's eye through the storyworld. An initial analysis beyond Homeric epic has yielded that the priming act is quite common in spontaneous spoken English narrative (see

---

51 Note the silence on the importance of this passage in the larger narrative in Geoffrey S. Kirk, *The* Iliad*: A Commentary* I. Cambridge 1985, 269; and Anton Bierl and Joachim Latacz, *Homers Ilias Gesammtkommentar* III.2. Berlin 2009, 22–26; Laura Slatkin, "Tragic Visualizing in the *Iliad*". In: Chris Kraus, Simon Goldhill, Helene P. Foley, and Jas Elsner (eds.), *Visualizing the Tragic: Drama, Myth, and Ritual in Greek Art and Literature, Essays in Honour of Froma Zeitlin*. Oxford 2007, 19–34 thematizes the importance of the gaze in guiding the narrative, though without discussing this specific scene.

52 *Iliad* 3.21, 3.30, 3.154, 5.95, 5.711, 7.17, 8.251, 11.248, 11.575, 11.581, 14.440, 16.419, 17.198, 18.222, 18.530, 21.49, 21.418; *Odyssey* 3.34, 8.272, 15.59, 17.492, 22.407, 24.232, 24.391; cf. Edwards, *Sound, Sense, and Rhythm*, 54.

53 Consider especially the movements toward the sphere of the Gods by way of a perception of characters on the battlefield in *Iliad* 5.711, 7.17, and 17.198.

54 Compare Foley's "economical exchange," in Foley, *How to Read*, 128, see note 4 above.

55 See note 37 above.

Chafe 1994), but also in Chaucer, the Grimms' *Märchen*, the *Chanson de Roland*, and Neapolitan folk tales.[56] These corpora know differing degrees of formulaicness, but what they have in common is that they were typically received by a listening audience. It is especially in this discourse situation that the priming act has its use.

The range of constructions containing priming acts, of which I have discussed two, suggests that they were an entrenched part of (narrative) language by the time of the Homeric performers. If such verses had become conventional enough, they may have served as landmarks in the event of the narration, equivalent to what Foley calls "boundary lines," "[separating] one narrative entity (scene, speech, or the like) from the next."[57] One kind of boundary line he distinguishes in Stolac performances is the "'position change' signal," which functions to "locate a character at or move a character to the next place of action."[58] This means that beyond the original cognitive and discursive function of establishing a transition from one contextual frame to the next, such verses may become a metanarrative signal to mark the beginning of a new scene, like the strongly formulaic verses that introduce or cap direct speech in Homeric epic.

An exploration of this syntactically anomalous construction in Homer from a discourse perspective demonstrates its potential in narration and performance. Although priming acts can indeed be described as "pendant nominatives" or "left-dislocated," such descriptions are neither sufficient nor productive. Priming acts fulfill the cognitive desire to hear about the "who" before the "what", a desire that is especially strong in—but not limited to—the context of performance. By establishing and using (groups of) characters as cognitive landmarks, the performer can guide his audience through an immense storyworld filled with countless protagonists. Studies of narrative tend to focus on content and structure, but the linguistic form is equally crucial and informative. Only with an eye for the smallest details can we figure out which elements of a discourse are instrumental in accommodating the audience's seemingly effortless understanding of a narrative. Homeric priming acts and the constructions that they partake in provide a unique insight in how narrative language accommodates the limitations and exploits the abilities of human cognition.

---

56 The data is gathered in "Mobility of the Mind's Eye", a paper presented at the *Poetics and Linguistics Association* conference in Heidelberg, August 2013, and currently in preparation for publication.

57 Foley, *How to Read*, 104–105. Bonifazi [this volume] adduces Sherzer's term "structuring" to describe this on-line process in epic performance.

58 *Ibid.*, 105.

References

Andersson, Theodore Murdock. *Early Epic Scenery: Homer, Virgil, and the Medieval Legacy.* Ithaca, NY 1976.

Arend, Walter. *Die typische Szenen bei Homer.* Berlin 1933.

Bakker, Egbert J. *Linguistics and Formulas in Homer: Scalarity and the Description of the Particle "per".* Amsterdam 1988.

Bakker, Egbert J. "Boundaries, Topics and the Structure of Discourse. An Investigation of the Ancient Greek Particle de". *Studies in Language* 17:2 (1993): 275–311.

Bakker, Egbert J. *Poetry in Speech: Orality and Homeric Discourse.* Ithaca, NY and London 1997.

Bertrand, Nicolas. *L'ordre des mots chez Homère: Structure informationnelle, localisation et progression du récit.* Unpublished thesis, Université de Paris-Sorbonne 2010.

Bierl, Anton and Joachim Latacz. *Homers* Ilias *Gesammtkommentar* III.2. Berlin 2009.

Bonifazi, Anna. "Memory and Visualization in Homeric Discourse Markers". In: E. Anne Mackay (ed.), *Orality, Literacy, Memory in the Ancient Greek and Roman World,* 35–64. Leiden 2008.

Bonifazi, Anna. *Homer's Versicolored Fabric. The Evocative Power of Ancient Greek Epic Word-Making.* Washington, DC 2012.

Brown, Gilian and George Yule. *Discourse Analysis.* Cambridge 1983.

Chafe, Wallace L. "Givenness, contrastiveness, definiteness, subjects, topics, and point of view". In: Charles N. Li (ed.), *Subject and Topic,* 25–55. New York, NY 1976.

Chafe, Wallace L. "The Flow of Thought and the Flow of Language". In: Talmy Givón (ed.), *Discourse and Syntax,* 159–182. New York, NY 1979.

Chafe, Wallace L. *The Pear Stories: Cognitive, Cultural, and Linguistic Aspects of Narrative Production.* Norwood, NJ 1980.

Chafe, Wallace L. *Discourse, Consciousness, and Time: The Flow and Displacement of Conscious Experience in Speaking and Writing.* Chicago, IL 1994.

Chantraine, Pierre. *Grammaire Homérique II: Syntaxe.* Paris 1963.

Clark, Matthew. *Out of Line. Homeric Composition Beyond the Hexameter.* Lanham, MD 1997.

Clay, Jenny Strauss. *Homer's Trojan Theater: Space, Vision, and Memory in the* Iliad. Cambridge 2011.

Collins, Christopher. *Poetics of the Mind's Eye: Literature and the Psychology of Imagination.* Philadelphia, PA 1991.

Dancygier, Barbara. *The Language of Stories: A Cognitive Approach.* Cambridge 2012.

Dancygier, Barbara and Eve Sweetser (eds.), *Viewpoint in Language.* Cambridge 2012.

Devine, Andrew M. and Laurence D. Stephens. *The Prosody of Greek Speech.* New York 1994.

Dik, Helma. *Word Order in Ancient Greek: a Pragmatic Account of Word Order Variation in Herodotus.* Amsterdam 1995.

Dik, Helma. *Word Order in Greek Tragic Dialogue.* Oxford 2007.

Edwards, Mark W. *Sound, Sense, and Rhythm: Listening to Greek and Latin Poetry.* Princeton, NJ 2002.

Emmott, Catherine. *Narrative Comprehension. A Discourse Perspective.* Oxford 1997.

Fillmore, Charles J. "The need for a frame semantics within linguistics". In: Hans Karlgren (ed.), *Statistical Methods in Linguistics* 12 (1976): 5–29.

Fludernik, Monika. *Towards a 'Natural' Narratology.* London 1996.

Foley, John Miles. *Homer's Traditional Art.* University Park, PA 1999.

Foley, John Miles. *How to 'Read' an 'Oral' Poem.* Chicago, IL 2002.

Fowler, Richard. "The Homeric Question". In: *The Cambridge Companion to Homer,* 220–232. Cambridge 2004.

Fraenkel, Eduard. "Kolon und Satz II. Beobachtungen zur Gliederung des antiken Satzes". In: *Nachrichten der Göttinger Gesellschaft der Wissenschaften, Phil.-hist. Klasse*, 1933, 319–354.

Geluykens, Ronald. *From Discourse Process to Grammatical Construction: On Left-Dislocation in English.* Amsterdam 1992.

Gentili, Bruno and Liana Lomiento. *Metrica e ritmica: storia delle forme poetiche nella Grecia antica.* Milan 2003.

Goldstein, David Michael. *Wackernagel's Law in Fifth-Century Greek.* Doctoral thesis, UC Berkeley 2010.

Hajnal, Ivo. "e-ke-qe oder e-ke?—Mündlichkeit und Schriftlichkeit in den mykenischen Palastarchiven". In: Thomas Krisch, Thomas Lindner, and Ulrich Müller (eds.), *Analecta Homini Universalis Dicata. Arbeiten zur Indogermanistik, Linguistik, Philologie, Politik, Musik und Dichtung. Festschrift für Oswald Panagl zum 65. Geburtstag,* 233–251. Stuttgart 2004.

Halliday, M.A.K. and Ruqaiya Hasan. *Cohesion in English.* London 1976.

Hannay, Mike and Caroline H. Kroon. "Acts and the Relationship Between Discourse and Grammar". *Functions of Language* 12 (2005): 87–124.

Havelock, Eric A. *Preface to Plato.* Cambridge, MA 1963.

Hellwig, Brigitte. *Raum und Zeit im homerischen Epos.* Tübingen 1964.

Hengeveld, Kees and J. Lachlan Mackenzie. *Functional Discourse Grammar. A Typology-Based Theory of Language Structure.* Oxford 2008.

Herman, David. *Story Logic. Problems and possibilities of narrative.* Lincoln 2002.

de Jong, Irene J.F. *Narrators and Focalizers. The presentation of the story in the* Iliad. London 2004, 2nd ed.

de Jong, Irene J.F. *Homerus:* Iliad *book XXII.* Cambridge 2012.

de Jong, Irene J.F. (ed.). *Space in Ancient Greek Literature.* Leiden 2012.

de Jong, Irene J.F., René Nünlist, and Angus M. Bowie (eds.). *Narrators, Narratees, and Narratives in Ancient Greek Literature.* Leiden 2004.

Jordan, Hedwig. *Der Erzählungsstil in den Kampfszenen der* Ilias. Breslau 1905.

Kirk, Geoffrey S. *The* Iliad*: A Commentary I-VI.* Cambridge 1985–1993.

Kroon, Caroline H. *Discourse Particles in Latin.* Amsterdam 1995.

Landau, Barbara and Ray Jackendoff. "'What' and 'Where' in Spatial Language and Cognition". *Behavioral and Brain Sciences* 16 (1993): 217–265.

Lord, Albert Bates. *The Singer of Tales.* Cambridge, MA 2002, 2nd ed.

Minchin, Elizabeth. *Homer and the Resources of Memory: Some Applications of Cognitive Theory to the* Iliad *and the* Odyssey. Oxford 2001.

Minchin, Elizabeth. "Spatial Memory and the Composition of the *Iliad*". In: Anne Mackay (ed.), *Orality, Literacy, Memory in the Ancient Greek and Roman World,*. 9–34. Leiden 2008.

Ochs Keenan, Elinor and Bambi Schieffelin. "A Reconsideration of Left Dislocation in Discourse". In: *Proceedings of the 2nd Annual Meeting of the Berkeley Linguistics Society*, 1976, 240–257.

Parry, Adam (ed.). *The Making of Homeric Verse. The Collected Papers of Milman Parry*, Oxford 1971.

Pekarek Doehler, Simona. "Emergent Grammar for All Practical Purposes: The on-Line Formating of Dislocated Constructions in French Conversation". In: Stefan Pfänder and Peter Auer (eds.), *Constructions: Emerging and Emergent,* 46–88. Berlin and New York 2011.

Purves, Alex C. *Space and Time in Ancient Greek Narrative*. Cambridge 2010.
Richardson, Scott. *The Homeric Narrator*. Nashville, TN 1990.
van Riemsdijk, Henk and Frans Zwarts. "Left dislocation in Dutch and the status of copying rules". In: Elena Anagnostopoulou et al. (eds), *Materials on Left Dislocation*, 13–31. Amsterdam 1997 [1974].
Roulet, Eddy. "Échanges, interventions et actes de langage dans la structure de la conversation". *Études de linguistique appliquée* 44 (1981) : 7–39.
Rubin, David C. *Memory in Oral Traditions: The Cognitive Psychology of Epic, Ballads, and Counting-Out Rhymes*. New York, NY 1995.
Ruijgh, Cornelis J. "La place des enclitiques dans l'ordre des mots chez Homère d'après la loi de Wackernagel". In: Heiner Eichner and Helmut Rix (eds.), *Sprachwissenschaft und Philologie: Jacob Wackernagel und die Indogermanistik heute*, 213–233. Wiesbaden 1990.
Ryan, Marie-Laure. "Cognitive Maps and the Construction of Narrative Space". in: David Herman (ed.), *Narrative Theory and the Cognitive Sciences*, 214–242. Stanford, CA 2003.
Salmon, William. "Double Subjects and Verbal Demonstrations". *Journal of Pragmatics* 42 (2010): 3431–3443.
Schank, Roger C. and Robert P. Abelson. *Scripts, Plans, Goals, and Understanding: An Inquiry into Human Knowledge Structures*. Hillsdale, NJ 1977.
Scheppers, Frank. *The Colon Hypothesis. Word order, discourse segmentation and discourse coherence in Ancient Greek*. Brussels 2011.
Sinclair, John McHardy and R. Malcolm Coulthard. *Towards an Analysis of Discourse: The English Used by Teachers and Pupils*. London 1975.
Slatkin, Laura. "Tragic Visualizing in the *Iliad*". In: Chris Kraus, Simon Goldhill, Helene P. Foley, and Jas Elsner (eds.), *Visualizing the Tragic: Drama, Myth, and Ritual in Greek Art and Literature, Essays in Honour of Froma Zeitlin*, 19–34. Oxford 2007.
Slings, Simon R. "Written and Spoken Language: an exercise in the pragmatics of the Greek sentence". *Classical Philology* 87 (1992): 95–109.
Taleb, Nassim Nicholas. *The Black Swan: The Impact of the Highly Improbable*. New York 2007.
Talmy, Leonard. "A Cognitive Framework for Narrative Structure". In: Leonard Talmy, *Toward a Cognitive Semantics* 2, 417–482. Cambridge, MA 2000.
Tsagalis, Christos. *From Listeners to Viewers: Space in the* Iliad. Washington, DC 2011.
Turner, Frederick and Ernst Pöppel. "The Neural Lyre: Poetic Meter, the Brain, and Time". *Poetry* 142 (1983): 277–309.
Turner, Mark. *The Literary Mind*. Oxford 1996.
Wackernagel, Jacob. "Über ein Gesetz der indogermanischen Wortstellung". *Indogermanische Forschungen* 1 (1892): 333–436.
Winograd, Eugene and Vaughan E. Church. "Role of Spatial Location in Learning Face-Name Associations". *Memory and Cognition* 16.1 (1988): 1–7.
Zoran, Gabriel. "Towards a Theory of Space in Narrative". *Poetics Today* 5.2 (1984): 309–335.

Sonja Zeman, LMU Munich

Orality, visualization, and the Historical Mind

The ‘visual present’ in (semi-)oral epic poems and its implications for a Theory of Cognitive Oral Poetics

Abstract: With the aim of bringing together oral poetics and cognitive linguistics, this chapter addresses the present tense as one of the most prominent patterns of both visualization and oral narratives. By examining present tense usage in Middle High German (semi)oral epic poems, it will be argued that its main function cannot be accounted for in terms of ‘vividness’ or ‘concreteness’. Rather, it lies within its meta-narrative usage as it simulates a complex relationship of simultaneity between narrator, audience, and the story world, and thus contributes to establishing a shared experience in the sense of a common ground. Through a comparison of the medieval and Homeric tradition, the analysis leads both to a specification of parameters for cross-linguistic comparisons of oral poetic texts and to significant modifications with respect to ‘orality’ and ‘visualization’ as key constructs for a Theory of Cognitive Oral Poetics.

1 ‘Visualization’ as a universal feature of oral epic poems

“I sing what I see”[1]

If one were to look at the literature of oral poetics from a cross-linguistic point of view and to search for the universal features of oral epic poems, ‘visualization’ would be a good aspect. One of the most basic characteristics of oral composition is generally understood to lie in its ‘vividness’ and ‘concreteness’. With respect to Old French epic poems, for example, Fleischman has emphasized their ‘optic quality’,[2] which she claims to be a linguistic reflex of the act of performance.

1 Suzanne Fleischman, *Tense and narrativity. From medieval performance to modern fiction*. London 1990, 265.

2 Ibid.; Suzanne Fleischman, “Evaluation in Narrative: The Present Tense in Medieval ‘Performed Stories’”. *Yale French Studies* 70 (1986): 199–251.

According to her view, it is the reception situation shared by the poet and a listening audience that leads to the fact that the epic poet sings of past events as if observing them simultaneously to his performance—a strategy which Fleischman has dubbed "I sing what I see": "The epic poet sings of past events but endeavors to report them as if observing them now [...]."[3]

Such visual quality has also been described as characteristic for Ancient Greek epic poems. The strategy of the Homeric narrator resembles the narrating mode in Old French texts insofar as the narrator "adopts the stance of an eyewitness"[4]. In this respect, the Homeric tradition has also been characterized as an "optic poetics":

> Vision, in fact, is the privileged sense in Homer in spite of its dependence on hearing (*kleos*); the poetics of the Homeric tradition is an optic poetics. The Homeric narrator typically adopts the stance of an eyewitness to the events he describes, [...] sometimes even staging his audience as spectators of the scene [...].[5]

The witness of "[t]he living eye of the singer's imagination"[6] as a narrative principle has been documented for other epic traditions as well.[7] The cross-linguistic view hence indicates that "oral traditions all over the world are characterized by vivid, concrete, visual detail".[8] While empirical evidence seems to show that 'visualization' has to be seen as a universal feature of oral epic poems, it is still disputed *why* oral composition should lead to 'visual' literary structures. With respect to a Theory of Cognitive Oral Poetics, this appears to be a question worth being pursued: If—as the empirical evidence suggests—textual patterns of visualization constitute a universal feature of oral poetry, their examination is meant to lead to the core principles of oral poetics.

Furthermore, 'visualization' has also been a key construct in investigations of cognitive linguistics. As cognitive linguistics is concerned with the general principles of linguistic-cognitive processes underlying the production and reception of language, imaginative structures like conceptual metaphors are central objects of

---

3 Fleischman, *Tense ad Narrativity*, 265.
4 Egbert J. Bakker, *Pointing at the past. From Formula to Performance in Homeric Poetics*. Cambridge, MA and London 2005, 63.
5 Ibid.
6 Albert Bates Lord, *The Singer of Tale*. Ed. by Stephen Mitchell and Gregory Nagy. Cambridge, MA [1960] 2000, 92.
7 See e.g. David F. Elmer, "Presentation Formulas in South Slavic Epic Song". *Oral Tradition* 24.1 (2009): 41–59, and Ruth Finnegan, *Oral literature in Africa*. Cambridge [1970] 2012.
8 Bakker, *Pointing*, 66; cf. also David C. Rubin, *Memory in oral traditions. The cognitive psychology of epic, ballads, and counting-out rhymes*. Oxford 1995, 8.

interest since mental images are seen as deeply linked with cognitive capacities of the human mind.

In this respect, patterns of textual visualization lie at the intersection of cognitive linguistics' and oral poetics' interests and thus constitute a promising object of investigation for addressing the question of how we can link both research areas together. With this aim in mind, this chapter addresses one of the most prominent patterns of both visualization and oral poetics: the use of the present tense, also dubbed "présent visuel".[9] Taking a typological view as a starting point, Section 2 demonstrates that the present tense usage varies both cross-linguistically and within the different diachronic stages of language. In order to address the question of its general relationship to oral poetry, Section 3 focuses on the present tense usage in a specific condition, namely its usage within the (semi-)oral epic poems of Middle High German (MHG). Unlike other oral traditions, where the present tense is often used to denote events on the plotline in the sense of a Historical Present (HP), the MHG present tense is restricted to its utilization in meta-narrative devices. As will be shown by means of a detailed analysis, these meta-narrative devices simulate a complex relationship of simultaneity between narrator, audience and the story world as the present tense usage covertly indicates the narrator's point of view and its evaluative relation with regard to the narrated events. In this respect, they contribute to establishing a "shared experience"[10] in the sense of a *common ground*, so that an explanation of the present tense in terms of 'vividness' and 'concreteness' breaks down.

By taking into account the historic-cultural background, the following sections show that such analysis of the present tense usage leads to further implications with respect to the concepts of 'orality' (Section 4) and 'visualization' (Section 5). Based on a comparison of the medieval and Homeric tradition, it is argued that cross-linguistic investigations of oral poetry have to factor in different dimensions of 'orality', distinguishing between the compositional and performative aspects of oral epic poetics. By considering the specific conditions of oral tradition in the particular languages, it will moreover become clear that meta-narrative devices cannot only be seen as a reflection of the speaker's attitude toward the narrated events, but also that they reveal a socio-cultural presupposition with regard to 'reality' as a concept which is historically variable.

Drawing together the methodological and theoretical implications with respect to a Theory of Cognitive Oral Poetics, Section 6 emphasizes the need for

9 Fleischman, "Evaluation in Narrative".
10 Egbert J. Bakker, "Discourse and Performance: Involvement, Visualization and 'Presence' in Homeric Poetry". *Classical Antiquity* 12.1 (1993): 1–29, 18.

typological investigations in order to pin down the universal features of oral poetry: As the analysis of the present tense shows, a cross-linguistic view does not only reveal the different dimensions of 'orality', 'visualization' and *common ground* as central parameters for cross-linguistic investigations of oral epic poetry, but is also able to approach the basic questions which stand in the center of a Theory of Cognitive Oral Poetics, namely whether—and how—this could provide insights into the Historical Mind.

2 Oral poetry and the present tense

> The speaker in using it [i.e. the Historical Present; SZ] steps outside the frame of history, visualizing and representing what happened in the past as if it were present before his eyes.[11]

The usage of the present tense denoting past events of narration is commonly seen as one of the "most prominent expressions"[12] in oral narratives and, at the same time, as the most common pattern of visualization in oral epic poems. The so-called Historical Present (HP) thus seems to confirm the strong link between 'orality' and 'visualization'. Traditionally, the HP is described as "visualizing and representing what happened in the past as if it were present before his [i.e. the narrator's; SZ] eyes".[13] Commonly, this visual effect of the HP is considered to dramatize the story "by making the audience feel as if they were present at the time of the experience, witnessing events as they occurred".[14] As a result, Fleischman has considered this kind of present tense usage to be "exclusively a phenomenon of orally performed narratives".[15] Similarly, the use of the present tense in alternation with past tenses has been described as an "oral pattern" by Wolfson and Fludernik,[16] while Tristram, in a study on Middle Irish texts, considers the HP

11 Otto Jespersen, *The philosophy of grammar*. Chicago 1924, 258.

12 Peter Koch and Wulf Oesterreicher, *Gesprochene Sprache in der Romania*. Berlin and New York 2011, 74–75.

13 Jespersen, *Philosophy of grammar*, 258.

14 Fleischman, "Evaluation in Narrative", 203.

15 Fleischman, *Tense and Narrativity*, 376 fn. 22.

16 Monika Fludernik, "The Historical Present Tense Yet Again. Tense Switching and Narrative Dynamics in Oral and Quasi-Oral Storytelling". *Text* 11 (1991): 365–398; Monika Fludernik, "The historical present tense in English literature: an oral pattern and its literary adaptation". *Language and Literature* 17 (1992): 77–107; Nessa Wolfson, *CHP. The conversational historical present in American English narrative*. Dordrecht 1982.

to be a general indicator of a text's degree of 'orality'.[17] In this spirit, it has even been suggested that the HP is a universal phenomenon of oral narration.[18]

This latter suggestion is, however, clearly contradicted by empirical facts: Although the HP is found in many languages of oral tradition (e.g. in Biblical Hebrew,[19] Ancient Greek,[20] Latin,[21] Old French,[22] Old Islandic,[23] Middle Irish,[24] and Middle English[25]), it is not documented in every oral tradition. The HP does, for example, not occur in Homeric epic poems[26] or Old English[27]. It is also missing within the Old High and Middle High German texts.[28] Hence, while the visual character of oral epic poems might be a universal feature of oral epic poems, it can be taken for granted that its formal devices certainly are not. Having a closer look at German, the link between HP and 'orality' becomes even more questionable. In the history of German, the use of the HP does not occur before the 13th century and has, since then, spread only slowly. A significant rise of its application was not documented until the 16th century in Early New High German (ENHG).[29] That is to say that it is primarily used in a period of time which is characterized by a strong trend to literacy, while within the earlier 'oral' stages, the HP is completely missing.

---

17 Hildegard L. C. Tristram, "Tense and Time in Early Irish Narrative". *Innsbrucker Beiträge zur Sprachwissenschaft: Vorträge und kleinere Schriften* 32 (1983): 5–37.

18 Koch and Oesterreicher, *Gesprochene Sprache*, 170.

19 Stefano Cotrozzi, *Expect the unexpected. Aspects of pragmatic foregrounding in Old Testament narratives*. New York 2010.

20 C. M. J. Sicking and P. Stork, "The grammar of the so-called Historical Present in Ancient Greek". In: Egbert J. Bakker (ed.), *Grammar as interpretation. Greek literature in its linguistic contexts*. Leiden and New York 1997, 131–168.

21 Carlotta Viti, "The non-literal use of tenses in Latin, with particular reference to the praesens historicum". *Revue de Linguistique Latin* 5 (2010): URL: http://www.zora.uzh.ch/43735/ (accessed April 25, 2015).

22 Fleischman, *Tense and Narrativity*.

23 Ulrike Sprenger, *Praesens historicum und Praeteritum in der altisländischen Saga*. Basel 1951.

24 Tristram, "Tense and Time".

25 Fludernik, "Historical present"; Fludernik, "Historical present yet again".

26 Egbert J. Bakker, "Verbal aspect and mimetic description in Thucydides". In: Egbert J. Bakker (ed.), *Grammar as interpretation. Greek literature in its linguistic contexts*. Leiden and New York 1997, 7–54; Sicking and Stork, "Historical present", 133.

27 J. M. Steadman, "The Origin of the Historical Present in English". *Studies in Philology* 14 (1917): 1–45, 12.

28 Hugo Herchenbach, *Das Präsens historicum im Mittelhochdeutschen* (Palaestra CIV). Berlin 1911.

29 Bruno Boezinger, *Das historische Präsens in der älteren deutschen Sprache* (Leland Stanford Junior University Publications. University Series). Palo Alto, CA 1912, 12, with reference to Wilhelm Wilmanns, *Deutsche Grammatik*, Volume III. Straßburg 1906, 96).

The use of the present tense is thus obviously not the same as described for the Romance languages by Fleischman.[30] Hence, one could suggest that the MHG epic poems show a very specific relationship to 'orality'. But because there are no concrete reasons to assume that the MHG epic poems are somehow "less oral" than e.g. the Romance ones, it appears to be more natural to call into doubt the plain link between 'orality' and the 'présent visuel'. In this respect, the analysis of the German present tense usage in Middle High German epic poems will reveal that its effects of 'visualization' are just surface phenomena of a deeper functional pattern which are not directly linked to 'orality', but to the pragmatic presupposition of the act of performance.

3 The present tense in Middle High German (semi-)oral epic poems

> [. . .] the present tense serves to establish the clearest link with the mind of a character that is possible in a literary text,[31]

In order to see what distinguishes the German present usage from the HP found in other epic oral poems, we first have to make explicit what actually characterizes a Historical Present: Traditionally, the HP is described as a present tense denoting past events. Such definition, however, covers a rather wide range of different uses of the present tense and hence does not give sufficient criterion which would be able to account for the specific usage of the present tense in older texts. To account for its characteristic usage, it has thus to be defined by further features:[32] The HP is used (a) in narrative contexts (b) in alternation with past tenses and (c) refers to events 'on-plot'[33]. In this narrow sense, the HP is absent from the MHG

30 Fleischman, *Tense and narrativity.*

31 Theo Damsteegt, "The Present Tense and Internal Focalization of Awareness". *Poetics Today* 26.1 (2005): 39–78, 51.

32 Cf. in detail Sonja Zeman, "Vergangenheit als Gegenwart? Zur Diachronie des Historischen Präsens". In: Petra Maria Vogel (ed.), *Sprachwandel und seine Reflexe im Nhd.* (Jahrbuch der Gesellschaft für Sprachgeschichte 4). Berlin and New York 2013, 235–255.

33 Fludernik, "Historical present yet again"; the distinction 'on-' vs. 'off-plot' refers to the distinction between the level of narrator (i.e. 'off-plot') and the level of the character (i.e. 'on-plot'). Generally speaking, the character level comprises the narrated events, while generic comments, evaluative devices etc. are attributed to the narrator's level.

texts, as it does not refer to a sequence of narrated events, but is restricted to the narrator's comments, cf. (1).

PATTERN A: non-referential, meta-linguistic comment ('off-plot')

(1) *wie* ***gevâhe*** *ich nû mîn sprechen an,*
'How do I now begin my speaking

daz ich den werden houbetman / Tristanden sô ***bereite***
that I prepare the dignified protagonist Tristan in such a way

ze sîner swertleite, / daz man ez gerne verneme?"
for his knightly accolade / that one would like to hear it willingly?'

[Gttfried von Straßburg, *Tristan*, 4591–4595][34]

In (1), the narrator is not *narrating* the events, but talking about the *representation* of the next event that will follow in the story. The comment thus serves a meta-narrative function as it refers to the construction process of the poem. Such usage of the present tense has also been labeled "author's present"[35] as it seems to 'occur where the poet emerges from behind the plot'.[36] Unlike the HP, the present tense is thus bound to the narrator level of the story and used 'off-plot'. In this respect, the present tense has no referential function with regard to the events within the story world,[37] but indicates that the narrator is commenting *about* the events. It thus foregrounds the relation between the narrator and the 'told', whereby the story is represented as originating from dynamic on-line manipulating before the eyes of the audience.

Example (2) shows another common pattern of present tense usage, where not the relationship between the narrator and the story, but between the narrator and the audience is foregrounded. Characteristic for this pattern are imperatives

---

34 Gottfried von Straßburg, *Tristan*. According to the text by Friedrich Ranke, ed. by Rüdiger Krohn, Vol. 1. Stuttgart 2001, 282.

35 Heinrich Hempel, "Vom 'Präsens Historicum' im Deutschen". In: Heinrich M. Heinrichs (ed.), *Kleine Schriften. Zur Vollendung seines 80. Lebensjahres am 27. August 1965*. Heidelberg [1937] 1966, 422–429, 425.

36 Herchenbach, *Präsens historicum*, 86–87: „so tritt es [i.e. das Präsens; SZ] gern an Stellen, wo der Dichter hinter der Handlung hervortritt"; needless to say that 'author' (Hempel, "Präsens Historicum") resp. 'poet' (Herchenbach, *Präsens Historicum*) should not be referred to the actual writer of the story, but to the narrator.

37 Fleischman, "Evaluation in narrative", 213.

addressed to the hearer and verbs of perception which presuppose that narrator and audience are simultaneously looking upon the narrated events:

PATTERN B: *nu / alsô* + verb of perception + present tense ('off-plot' > 'on-plot')

(2) ***nû lât*** *Terramêren rîten –*
'**now let** Terramer ride –

hoeret *wie die êrsten strîten!*
Listen how the first ones are fighting!

sîn helfe ***kumt*** *in doch ze fruo.*
his help **comes** for them yet to early.'

[Wofram von Eschenbach, *Willehalm*, 360,29–361,1][38]

Examples of pattern B are often introduced by *nû* ('now') which is used here as a space builder. It indicates a new 'story now' and directs the attention to the shifted scene which is represented as co-temporal with the deictic frame of the recipient. Like in example (1), the narrative process is also seen as a constructive, dynamic one. In addition, the audience is directly addressed by means of imperatives (*lât*, *hoeret*) which simulate a perceptual simultaneity of hearing and seeing. This also holds for the present tense *kumpt* ('comes'), which seems to indicate an event of the story. Yet, it is still the narrator speaking about the events, slightly gliding back into the story world spread out before the hearer's imaginative eye. The present tense is thus used in a textual interface between 'on-' and 'off-plot' and marks the shift from one scene to another.

In (3), the present tense actually seems to refer to an event 'on-plot' as well. However, the present tense is also used in isolation and is situated at a potential interface between plotline and off-plotline as it indicates a new 'story now'. Like in the examples above, the stance of the narrator is presupposed. This is supported by the deictic particle *nu* ('now') which indicates that the actual 'story now' is represented as simultaneous with the time of the narrator and the audience, simulating a shared situation of 'viewing' the narrated events. The 'story now' and the act of narrating thus seem to be virtually conflated:

38 Wolfram von Eschenbach, *Willehalm*. Nach der Handschrift 857 der Stiftsbibliothek St. Gallen, ed. by Joachim Heinzle. Tübingen 1994, 327.

PATTERN C: *nu* / *alsô* + present ('on-plot')

(3) ***nu rennt*** *der künig Valerîn und Lanzelet ein ander an.*
'Now are riding the king Valerîn and Lancelot at each other.'

[Ulrich von Zatzikhoven, *Lancelet*, 5268][39]

In MHG, the present tense is thus restricted to the level of the narrator while the preterite denotes the plotline.[40] The general presupposition of the narrator level also becomes graspable in (4) as discussed by Herchenbach:

(4) *an dem* ***ervert nu*** *Parzivâl / diu verholnen maere umben grâl.*
'By him **gets to know now** Perceval the secret lore about the grail.'

[Wolfram von Eschenbach, *Parzival*, 452,29 f.][41]

Again, the present tense seems, at first glance, to refer to an event on the plotline. Herchenbach, however, suggests the following translation, cf. (4'):[42]

(4') *Von dem* ***sollte*** *Parzival jetzt die Kunde über den Gral* ***erfahren.***

'By him Perceval **was to get to know** [literally: '**should** get to know'] now the secret lore about the Grail.'

The preterite form of the modal verb *sollen* ('shall') + inf. in the German translation in (4') might initially seem surprising, as it does not denote a present time simultaneous to the time of the narrating subject, but an event that is, on the one hand, past with respect to the narrator and, on the other hand, future with respect to the protagonist. Nevertheless, the modal verb construction *sollte erfahren* adequately corresponds to the narrative function of the MHG present tense insofar as it indicates a split between the narrator level and the plotline of events. This is seen by the fact that two different perspectives arise, reflected by a different state of knowledge of the narrator and the protagonist: While the narrator in (4') knows what will happen later in the story, the protagonist on the event level (i.e.

39 Quoted in Herchenbach, *Präsens historicum*, 96.
40 Cf. in detail Sonja Zeman, *Tempus und "Mündlichkeit" im Mittelhochdeutschen. Zur Interdependenz grammatischer Perspektivensetzung und "Historischer Mündlichkeit" im mittelhochdeutschen Tempussystem*. Berlin and New York 2010.
41 Quoted in Herchenbach, *Präsens historicum*, 69.
42 Ibid., 70.

Perceval) is unaware of the events that will follow after the ‘story now’. Both present tense and the modal verb construction hence indicate that the narrator talks *about* the event to happen, but does not refer to the event itself.

Altogether, the usage of the present tense can thus be described by three patterns.[43]

PATTERN A **Metalinguistic, non-referential comment of the narrator**
→ **‘now’ of the narrating process**

PATTERN B ***nu / also* + simulated perception process**
(verbs of perception, interjections, imperatives)
→ **‘now’ of the perception process shared by narrator and audience**

PATTERN C ***nu / also* + event description**
→ **‘now’ of the ‘story world’**
(as synchronized with the perception process)

All three patterns have in common that they simulate a simultaneity between the ‘story now’, the narrator’s act of performance and the listening audience. This simultaneity, however, is not so much a temporal relation but, more importantly, tied to the structural difference between the level of the narrator and the level of the events. In this respect, the present tense functions as a meta-linguistic device insofar that it “establishes the discourse as something other than narration”.[44] Tense usage in MHG is thus quite regular.[45] However, it is not so much linked to a temporal opposition between ‘past’ vs. ‘non-past’, but reflects a conceptual opposition with regard to the speaker’s origo.[46] While the past tenses indicate a reference system which is distinct from the origo, the present tense always

43 Cf. also Zeman, “Vergangenheit als Gegenwart?”.
44 Fleischman, *Tense & narrativity*, 306.
45 This is worth particular mentioning as Fleischman, *Tense & narrativity*, considers ‘tense confusion’ as a general characteristic for the Old French texts. The regularity of the tense system in MHG however calls into question whether tense confusion is a universal feature of Oral Poetics.
46 In the tradition of Bühler, the notion *origo* is defined as an abstract zero point within a coordinate system which is as default constituted by the speaker’s ‘here and now’, but can be shifted in the sense of ‘Deixis am Phantasma’ (cf. Karl Bühler, *Sprachtheorie. Die Darstellungsfunktion der Sprache*. Stuttgart [1934] Reprint 1999). In this respect, two different basic constellations can be distinguished: The speaker’s origo can either be conceptualized as within its region of referential concern (‘origo-inclusive’)—or not (‘origo-exclusive’) (cf. also Gabriele Diewald, *Deixis und Textsorten im Deutschen*. Tübingen 1991).

presupposes an origo-inclusive deictic frame.[47] In consequence, the present tense induces an epistemic meaning in presupposing the speaker. Thus, the present tense can be seen as a reflection of a speaker's vantage point, and "not merely as rhetorical device [. . .] for an effect of immediacy".[48] Its inherent epistemic meaning enables its meta-linguistic use in foregrounding the level of the narrator and its occurrence where there can be taken for granted a shared pragmatic situation.

Whether pattern C constitutes the grammaticalization source[49] for the later HP will be a matter for future research. Nevertheless, the analysis above offers clear indications for such a view. With respect to MHG, Boezinger already expressed the assumption that the early occurrences of the present tense in the sense described above are meant to be seen as the precursors of the later HP.[50]

Further evidence for such line of development is taken from the fact that the HP has also been described as an evaluative device.[51] Such description is also compatible with the approach by Langacker, who characterizes the HP as kind of a "mental replay, where previous events are relived by running the tapes of memory".[52] In this respect, the HP does not refer directly to the original, actual events themselves but to their representations. This also suggests that the grammaticalization source of the HP has to be seen in the evaluative meta-narrative function of the present.

Against this background, we can thus conclude that the present tense usage in MHG is quite regular as it functions as a meta-linguistic device that foregrounds the level of the narrator by presupposing an origo-internal deictic reference frame. In this respect, the visual effects evoked by meta-narrative devices cannot be seen as plain representations of the singers' mental images, but are linked to a deeper functional pattern as they serve for establishing a *common ground* between the narrator and the audience. Thus, neither terms of 'concreteness' nor

---

47 Sonja Zeman, "More than 'time': The grammaticalization of the German tense system and the 'frame of reference' as a crucial interface between space and time". In: Kasia Jaszczolt and Luna Filipović (eds.), *Space and Time in Languages and Cultures. Language diversity.* Amsterdam and Philadelphia 2012, 157–178.

48 Tomoko Sakita, *Reporting Discourse, Tense, and Cognition.* Amsterdam 2002, 79.

49 Cf. Elisabeth Leiss, *Artikel und Aspekt. Die grammatischen Muster von Definitheit.* Berlin and New York 2000 with respect to the premise that the HP is not merely a stylistic device, but a grammatical pattern.

50 Bruno Boezinger, *Historisches Präsens*, 33–34.

51 Cf. e.g. Fleischman, "Evaluation in narrative"; idem, *Tense & narrativity*; Fludernik, "Historical present"; Sakita, *Reporting Discourse*; Cotrozzi, *Pragmatic Foregrounding.*

52 Ronald W. Langacker, "The English present. Temporal coincidence vs. epistemic immediacy". In: Frank Brisard and Adeline Patard (eds.), *Cognitive Approaches to Tense, Aspect, and Epistemic Modality.* Amsterdam and Philadelphia 2011, 45–86, 61.

'vividness' are able to account for its usage in MHG epic poems. So what does this mean with respect to the optic character of the present tense and its status in oral poetics? In other words: How 'oral' and how 'visual' is the 'visual present' actually? These questions are addressed in the following sections, claiming that it is the pragmatic presupposition of the performative aspect—and not so much 'orality' itself—which leaves its traces in the text.

4 Composition and performance: How 'oral' is the 'visual present'?

> There can be no poem without a performer to give it being [. . .][53]

'Oral poetry', in the sense of Lord's and Parry's original definition, refers to the fact that poems are orally composed in an immediate presence of the audience. It is thus the performative aspect of the poems' production which is seen as its relevant characteristic—that means the fact that an oral poem is "not composed *for* but *in* performance":[54]

> What is important is not the oral performance but rather the composition *during* oral performance.[55]

For the South Slavic Songs investigated by Parry, this seems at first to be a useful definition. With respect to historical stages of language, however, we have to tackle the methodological problem that the 'oral' poems are only known *because* they are transmitted in written records. The matter gets even more complicated when looking at MHG epic poems which already are grounded in a literary tradition: Unlike in Homeric poetry, the reference point for composing an epic poem in MHG is the written word as the singers' research for suitable subjects and their historical details is based on books. This is obvious by the fact that the narrators emphasize their elaborate literature research in order to present the

---

53 Mark C. Amodio, *Writing the oral tradition. Oral Poetics and Literate Culture in Medieval England.* Notre Dame, IN 2004, xvi.
54 Lord, *Singer of Tale*, 13.
55 Ibid., 5.

'right' version of their story.[56] Whether these assertions reflect the real creative process or not, it is clear that the source of storytelling is not seen in the muses' afflatus but the knowledge of the written word.[57]

With respect to the MHG epic poems, literacy and orality are thus intricately intertwined as seen in their textual structure. While the story is claimed to be based on the reference of written sources, the meta-narrative devices—as illustrated before—refer to a performative situation in addressing a listening audience. Even if this cannot—without reservations—be considered a direct reflection of oral performance because the epic poems in MHG could both be presented orally and read as texts, it is known that free oral presentation (i.e. without being based on any written material) is rather the rule than the exception.[58] MHG epic poems are thus rooted both in oral tradition and the written (i.e. Latin) word. Due to this complex relationship between 'orality' and 'literacy'—Däumer speaks of a 'third kind of medial condition' which involves a specific semiotic premise —,[59] the MHG epic poems cannot be seen as 'oral poetry' in a strict sense of the term, i.e. as poems "composed in a manner evolved over many generations by singers of tales who did not know how to write".[60] In consequence, MHG epic poems have been labeled 'semi-oral' or 'transitional texts'. Such terminology, however, is problematic as it presupposes a straight line of development from 'orality' to 'literacy' without considering their mutual interdependencies against the historical background. In order to account for this complex relationship, we thus have to distinguish the different *kinds* of 'orality' which are hidden behind that term.

As seen above, 'oral' within the Parry-Lord-tradition has been defined in a technical sense, that is to say, it depends on the premise that 'orality' is linked to the medial dichotomy between 'oral' and 'written' which is seen as a mutually exclusive opposition.[61] In this respect, the features of oral epic poems are seen as reflections of the incremental processing in oral communication: formulae and repetitions are attributed to the fact that the poems are composed 'on line' (in the

56 Cf. e.g. the narrator's prolog in Gottfried von Straßburg's *Tristan* (131–172), which ensures that he will tell the story *rehte* ('rightly', 134) by referring to the 'right' literary source, namely the "well read" Thômas von Britanje.

57 Cf. Harald Haferland, *Mündlichkeit, Gedächtnis und Medialität. Heldendichtung im deutschen Mittelalter*. Göttingen 2004, 66 with respect to the fact that the concept of writing is not mentioned at all in the Homeric poems.

58 Cf. Harald Haferland, *Mündlichkeit*, 463; Matthias Däumer, *Stimme im Raum und Bühne im Kopf. Über das performative Potenzial der höfischen Artusromane*. Bielefeld 2013.

59 Däumer, *Stimme im Raum*.

60 Lord, *Singer of Tale*, 4.

61 See also Bakker, *Pointing*, 44.

sense of Auer)[62], i.e. simultaneously to their reception, and thus are conditioned by cognitive parsing restrictions. Such definition of 'orality' thus refers to the medial dimension of 'orality'. It fails, however, to account for 'oral' poems surviving in written texts, as a text can more or less reflect an 'oral style', irrespective of its actual medial transmission. In this sense, 'cognitive orality'[63] resp. 'conceptual orality'[64] is a gradual concept independent of the medial dichotomic bias between 'spoken' vs. 'written', and hence allows for talking about 'orality' in written texts.

'Historical orality'—i.e. the dimension of an 'oral style' that is attributed to historical stages of language and hence preserved only indirectly in *written* texts —thus cannot be anything else than 'conceptual/cognitive orality', as there is no methodical access to medial orality. Yet, 'conceptual orality' remains a vague concept as it is still an open question concerning what features are actually meant to characterize an 'oral style'.[65] With regard to 'historical orality', this issue is particularly questionable as the socio-cultural condition of 'orality' in a particular linguistic community and its degree of literacy is not a constant one but historically variable. In this respect, texts of 'oral cultures' have been claimed to be 'more oral' in general: It is assumed that their 'oral ductus' is inscribed within the text as a presupposition, as "the traditional oral register remains a powerful and flexible significative idiom, whether it is produced by the mouth or the pen".[66]

While 'cognitive/conceptual orality' seems, at first glance, to be a suitable term for addressing 'historical orality', it nevertheless fails to account for the differences between the particular oral phenomena which can be traced back to different aspects of orality: While the formulaic style and the repetitions in oral epic poems have been considered to be reflections of the poem's on line processing, we have seen earlier that the 'visual present' is not so much the consequence of the text's production, but its communicative transmission in oral presentation. In other words, we have to factor in a third distinction, namely between *composition* and *performance*. While the fact that oral poems are produced 'on line' entails textual features which are directly linked to the (medially) oral situation in a

62 Peter Auer, "On line-Syntax—Oder: was es bedeuten könnte, die Zeitlichkeit der mündlichen Sprache ernst zu nehmen". *Sprache und Literatur* 85 (2000): 43–56.

63 Fleischman, *Tense and Narrativity*, 89.

64 Peter Koch and Wulf Oesterreicher, "Sprache der Nähe – Sprache der Distanz. Mündlichkeit und Schriftlichkeit im Spannungsfeld von Sprachtheorie und Sprachgeschichte". *Romanistisches Jahrbuch* 36 (1985): 15–43.

65 Cf. for discussion Mathilde Hennig, *Nähe und Distanzierung. Verschriftlichung und Reorganisation des Nähebereichs im Neuhochdeutschen.* Kassel 2009, and Sonja Zeman, "Historische Mündlichkeit. Empirische Erörterung einer theoretischen Problemlage". *Zeitschrift für Germanistische Linguistik* 41.3 (2013): 1–36.

66 Amodio, *Writing*, 42.

technical sense, other 'oral' characteristics can be traced back to the pragmatic conditions of a performance situation—respectively: to the *simulation* of a performance situation, as it becomes more and more the case in later medieval texts. Although the distinction between composition and performance is crucial for every kind of oral poetics, the need for its differentiation becomes particularly clear in the medieval context where, with the transition to the recitation of written literature, the features of composition are beginning to lose their motivation whereas the style of oral performance persists, as laid out by Fleischman:

> In written texts that were still read aloud (e.g., chronicles and prose romances), the mechanical features relating to *performance* were still functionally motivated, whereas those relating to oral *composition* had lost their mechanical raison d'être.[67]

Against this background, we can thus distinguish the following dimensions of 'orality': Besides the difference between *medial* and *conceptual/cognitive orality*—the former referring to 'orality' in the technical sense of oral processing of spoken words, the latter addressing the general notion of an 'oral style' independent from the medial bias between 'spoken' and 'written'—'conceptual orality' for historical stages has to be further differentiated with respect to the distinction of composition vs. performance. Thereby, compositional aspects of oral epic poems—although only preserved in written texts—can be seen as reflections of medial conditions of orality as its features like formula, repetitions, structures of verb-serialization etc. can be traced back to cognitive parsing restrictions linked to the production process of on line processing. The characteristics of performance, on the other hand, are not primarily triggered by the medial aspects of orality, but by the pragmatic presupposition of a shared communicative space of singer and audience which is indicated by origo-deictic expressions. Hence, these features are not primarily 'oral' in the sense of 'medial orality', but linked to the concept of 'involvement'[68] and 'proximity'.[69] In other words: The statement that 'historical

---

67 Fleischman, *Tense and Narrativity*, 88.

68 Wallace Chafe, "Integration and Involvement in Speaking, Writing, and Oral Literature". In: Deborah Tannen (ed.), *Spoken and Written Language: Exploring Orality and Literacy*. Norwood 1982, 35–53. With respect to the conceptual difference between Chafe's 'involvement' and 'Nähesprache' ('language of proximity'; Koch and Oesterreicher, "Sprache der Nähe", and Vilmos Àgel and Mathilde Hennig (eds.), *Grammatik aus Nähe und Distanz. Theorie und Praxis am Beispiel von Nähetexten 1650–2000*. Tübingen 2006) see Zeman, "Historische Mündlichkeit"; Sonja Zeman, "Nähe, Distanz und (Historische) Pragmatik. Oder: Wie 'nah' ist 'Nähesprache'?". In: Mathilde Hennig and Helmuth Feilke (eds.), Zur Karriere von 'Nähe und Distanz'. Berlin and Boston), 263–302.

69 Note that, although 'involvement' (and 'Nähesprache') are discussed under the label of 'orality', its characteristic features cannot be derived from the medial condition of oral commuorality' (i.e. 'orality' as preserved in *textual* sources) can be by definition nothing other than 'conceptual' resp. 'cognitive orality' is certainly true. It is, however, a trivial fact that not only does not allow for any further insights into the very characteristics of 'historical orality', but, what is more, also prevents us from seeing the different factors triggering different features of 'orality'. In order to investigate 'orality' in historical stages of language and develop an adequate approach of 'oral style', thus different levels of orality have to be distinguished, which leave their traces in the text in different ways, cf. Fig. 1:

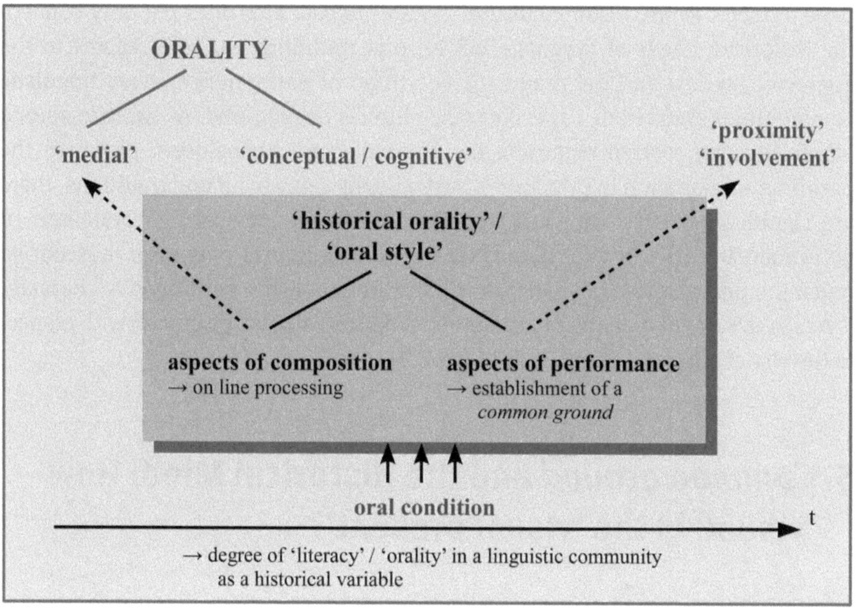

Figure 1: Dimensions of the macro-category 'orality'

How does the 'visual present' fit into this model? As seen in Section 3, the MHG present tense activates an origo-internal deictic reference frame. With respect to the double-layered structure of narrative discourse, its usage presupposes the narrator's point of view. Against this background, the affinity between present tense and 'orality' appears to be natural: Like other deictic devices which indicate a simultaneous relationship between the singer, the audience, and the narrated story, the present tense usage is linked to the performative character of oral

nication *per se* (cf. Zeman, "Historische Mündlichkeit"; Sonja Zeman, "Nähe, Distanz und (Historische) Pragmatik".

narration. It is, in contrast, not a feature of oral composition in the technical sense of the term. The meta-narrative devices are thus not a reflex of the medial opposition between 'oral' and 'written', but are linked to the performative character presupposed in the narrative setting.

In order to lay the methodological ground for differentiating between historical-specific and universal features of oral poetics, it is thus obvious that a simple binary distinction between 'written' vs. 'spoken' is not sufficient to account for the characteristics of 'oral' epic poems. Besides the need for a more fine-grained model which integrates the different dimensions of 'orality', it is also clear that we have to factor in the historic-cultural circumstances. This does not only concern the historical stages of language but all oral traditions as, with respect to the degree of literacy and the pragmatic condition of performance, every linguistic community is situated in a specific socio-cultural background. As Amodio among others has emphasized regarding the pragmatic presuppositions: Although the performance situation in principle is basically the same in all oral traditions, there are significant differences with respect to the language-specific parameters of performance.[70] The crucial parameter for analyzing oral poetics is thus not so much the performative character itself but rather *how* the performative character is realized. As shown in the following section, this will also be crucial with respect to the visual effects of the present tense usage.

5 *Common ground* and the Historical Mind: How 'visual' is the 'visual present'?

> The poet's "seeing" of reality is a remembrance of earlier poets'—and ultimately the Muses'—seeing of that same reality.[71]

Traditionally, the use of the present tense has been described in visual terms, as it seems to 'visualize' the past before the hearer's eyes. In this respect, it seems to trigger a "*quasi-perceptual experience*", as it has been described for mental imagery that occurs in the absence of external stimuli. In Section 3, however, it was argued that the visualizing effect of the present tense is merely a surface phenomenon of its deeper deictic function. As stated before, the narrator's com-

70 Amodio, *Writing*, 42.
71 Bakker, *Pointing*, 70.

ments are more than mere 'vivid' pictures of the represented scene placed in the 'mind's eye' of the listener. What has been shown to be crucial, however, is that they draw the attention to the actual communicative situation, conceptualized as a shared space between the poet and his audience—in other words: to the actual moment of the situation of performance. In this respect, the narrator's insertions in the present tense could be considered as a reflection of the communicative presuppositions of oral story telling.

But there is more than just the involvement of the audience and the shared experience within the actual performance in the sense of a *common ground* 1. As Elmer—drawing to Benveniste's distinction between *histoire* and *discours*[72]—has illustrated for South Slavic epic poems, narrator comments, such as presentation formula, presuppose the discourse mode which relies on the interaction of a speaker and a listener. Characters and events can thus "be presented—can be made present—to the audience because 'here' and 'now' are features of the performance, and not of the spatial or temporal framework of the narrated events".[73] The narrator's insertions in the present tense function in a very similar way as they are linked to the discours mode and presuppose an origo-internal reference frame. What the hearer 'sees' is hence not so much the events themselves, but the narrator's representation of these events. In Wagner's words, "[t]he narrator forces his audience to observe him observing the narration".[74] It is in this meta-narrative function that the relation between the narrator and the story comes to the fore—using Parry's words: the relation between *individual* and *tradition*.[75] As is commonly known, the stories underlying oral epic poems are "not themselves the original fictions of certain authors, but creations of a whole people, passed through one generation to another".[76] Therefore, the reference to the events is not merely an idea in the poet's mind but "an activation of tradition".[77]

---

72 Émile Benveniste, "Les relations de temps dans le verbe français". *Bulletin de la Société de Linguistique de Paris* 54 (1959): 69–82.

73 Elmer, "Presentation formulas", 48.

74 Silvan Wagner, "Narrator and narrative space in Middle High German Epic Poetry". In: Natalia Igl and Sonja Zeman (eds.), *Perspectives on narrativity and narrative perspectivization.* Amsterdam and Philadelphia, 115–138.

75 Milman Parry, "A comparative study of diction as one of the elements of style in Early Greek Epic poetry". In: Adam Parry (ed.), *The making of Homeric Verse: The Collected Papers of Milman Parry.* Oxford 1987, 421–436.

76 Ibid., 421.

77 Bakker, "Discourse and performance", 20.

In this respect, oral epic poems are closely linked to memory, which has often been emphasized in literature.[78] Besides the technical aspect of recollection in oral performance which is seen as the factor triggering mnemotechnic means such as formula, repetitions etc. and affects the aspect of oral composition (cf. Section 4), the basic function of oral epic poems lies in the reactivation and reassurance of a speech community's collective memory. In this wider sense, oral poetry as a whole is nothing other than establishing a common ground—not only with respect to the actual performance situation (*common ground* 1), but also with respect to the traditional basis underlying the process of storytelling (*common ground* 2). This becomes particularly clear in the Homeric formulae which have been considered as condensed expressions of cultural knowledge as they present the world "as all men (all men within the poem, that is) commonly receive it":[79]

> The style of Homer emphasizes constantly the accepted attitude toward each thing in the world, and this makes for a great unity of experience.[80]

While the function of preserving collective knowledge seems to be a crucial factor in all oral traditions, the presuppositions for this oral storage of cultural information are essentially different. This becomes evident when comparing the Homeric and the medieval tradition, both relying on different conceptions of 'memory'. For the Homeric poet, the act of reciting the story is an active process of construal: He constructs the story before the readers' eyes by unfolding the picture as seen in his mind, inspired by the muses' afflatus. As the construal of the story is a "cognitive act of re-creation",[81] the story itself is not seen as an externally given fact, but as dependent on the poet's perception:

> As in Berkeley's metaphysical system, there is in epic no objective reality independent of the narrator (the epic singer), for the epic past exists only as perception, both in the memory of the singer and the imagination of his audience.[82]

78 Bakker, "Discourse and performance"; Bakker, "Verbal aspect"; Bakker, *Pointing*; Rubin, *Memory*; Maria Dobozy, *Re-Membering the Present. The Medieval German Poet-Minstrel in Cultural Context*. Turnhout 2005; Mary Carruthers, *The book of memory: a study of memory in medieval culture*, Cambridge [1990] 2008.
79 Parry, "Language", 3.
80 Ibid.
81 Egbert J. Bakker, "*Mimesis* as Performance: Rereading Auerbach's First Chapter". *Poetics Today* 20.1 (1999): 11–26, 24.
82 Kawashima, "Narrative perspective", 114.

The epic singers are thus operative authorities of epic evidence who are allowed to "reactivate the epic reality *now*, in the context of the performance".[83] In this sense, the act of singing is really an act of 'seeing'.

As pointed out above, medieval epic poems also refer to the construction process of the story. However, the situation is different as memory has already begun to be a *thesaurus*[84], a "storage space 'from which' we take or 'where' we put something, to preserve it or to retrieve it".[85] Telling the story is thus not merely a cognitive act of perception, but rather an act of the story's perspectivization which is given—external to the mind—as a historical fact. In consequence, the responsibility for the poet is not so much to activate the story in the present performative situation, but to give the 'right' version of the story (cf. Section 4 above and Fn. 56). Thus—similar to the prose narrator in Ancient Greek—the poet "*fictionalizes* the act of perception, in order to represent the reality described".[86] That means, although the story is presented in an actual performance situation, it is clear that the story is taking place in 'the past' resp. in a narrative space distant from the actual performance situation.

This also becomes obvious by a look at the later development of meta-narrative comments in the fourteenth century prose novel which is, in contrast to the earlier epic poems, detached from performative mediation, cf. (5):

(5) ***Nu múßen ir diße rede laßen*** *von mym herren Gawan und von synen gesellen und* ***sprechen furter*** *von dem konig Artis ein wile.*

'**Now we have to leave** this talk about my lord Gawain and his companions and **speak further** a while about king Arthur.'

[*Prosa-Lancelot*, 25a,10][87]

In comparison with the narrator's comments in the epic poem analyzed in Section 3, the difference lies within the fact that there is not a single space which is represented as conflating the story world and the space of performance, but different narrated spaces which are distinct from the narrator's reference system

83 Bakker, *Pointing*, 64.
84 Carruthers, *Book of memory*, 37.
85 Bakker, *Pointing*, 140.
86 Bakker, "Mimesis", 21; emphasis in original.
87 *Prosalancelot II. Lancelot und Ginover II.* Nach der Heidelberger Handschrift Cod. Pal. germ. 147, herausgegeben von Reinhold Kluge, ergänzt durch die Handschrift Ms. allem. 8017–8020 der Bibliothèque de l'Arsenal Paris, Translated, commented, and edited by Hans-Hugo Steinhoff. Frankfurt am Main 1995.

and can be foregrounded in alternation by the narrator. An equivalent pattern for scene shift has also been described for Middle English by Fludernik, who suggests that this "standard pattern" for indicating a scene shift is a functional pattern deeply linked to the fact that the oral tradition has faded and the narrator has become a mediator between the book and his audience.[88]

The comparison between Homeric epic poems and the medieval tradition thus shows that both are 'visual' in a very wide sense of the term as they both contain mental imagery. Yet it has also become obvious that there are crucial differences linked to the presuppositions of the historic-cultural condition. Against this background, it becomes clear that 'memory', 'orality', and 'visualization' are not steady but historically variable concepts. As Kawashima has pointed out, this also holds for the concept of 'reality' itself. In Homeric tradition, as seen above, the space of performance, as a situation shared by the singer and the audience, and the narrated world are blended together in one single space. As a consequence, the entities of the narrated world functionally exist only while they are perceived by the singer.[89] Thus, the oral epic poems are "reality resulting from speech".[90] Such a narrative 'monism' is clearly contrasted with a concept of reality lying external to the mind, and which is closely linked to concepts such as 'literacy', 'historicity', and 'fictionality'. In this respect, oral epic poems rely on a shared epistemological *common ground* 3, which comprises general epistemological presuppositions concerning the speaker's relationship to what is conceptualized as 'reality' as shared knowledge within a speech community. In this sense, *common ground* 3 is not reflected by any linguistic means but becomes clear by cross-linguistic comparisons, as seen above. Based on the considerations so far, we can thus conclude that different dimensions of *common ground* are decisive factors in order to analyze oral poetry as they are crucial with regard to account for the complex relationship between the singer, his audience and the recounted:

Common ground$_1$: **→ Communicative situation ('singer' ↔ 'recipient')**
(e.g. the deictic reference frame shared by epic singer and audience)

Common ground$_2$: **→ Socio-pragmatic condition ('individual' ↔ 'tradition')**
(e.g. the world knowledge shared by the speaker community)

---

88 Monika Fludernik, "The diachronization of narratology". *Narrative* 11.3 (2003): 331–348, 335 and 338.

89 Robert S. Kawashima, "What is Narrative Perspective? A Non-historicist Answer". In: Robert S. Kawashima, Gilles Philippe, and Thelma Sowley (eds.), *Phantom sentences. Essays in linguistics and literature presented to Ann Banfield*. Bern and New York 2008, 105–126, 108.

90 Bakker, "Discourse and performance", 20.

Common ground$_3$: → **Epistemological presupposition ('speaker' ↔ 'reality')**
(e.g. the presupposition concerning the status of 'reality')

It is thus not 'vividness' or 'concreteness', but the involvement of the audience in a *common ground* that accounts for the 'visual' effects of the present tense. While this aspect seems to be crucial for all oral traditions, the differences lie in *how* the *common ground* is linked to different structures of 'visualization'—and what aspects are triggered by the presupposition of 'seeing reality'. While I will have to leave such epistemological questions for future research, it has nevertheless become clear that cross-linguistic comparisons of epic oral poems are indeed a suitable tool for gaining insight into the Historical Mind.

6 Theoretical and methodological implications for a Theory of Cognitive Oral Poetics

> 'Truth,' a word that is used frequently in oral performances, apparently refers not to an ontological or referential concept of truth, but to 'validity,' the validation of the stance of the story by bringing it into line with the stance of its audience.[91]

Where have the previous considerations led us so far? Starting point has been the empirical observation that there seems to be a strong link between 'visualization' and 'orality', which commonly has been traced back to the simultaneity of production and perception in oral performance. The present tense has been described as invoking narrative 'tableaus', 'visualizing' the narrative, and making it more 'vivid'. A closer look from a cross-linguistic perspective, however, has revealed that the interdependencies are far from being that simple. As it has been shown, the present tense usage in MHG oral epic poems can neither be accounted for in terms of 'visualization' or 'vividness', nor is this feature restricted to 'orality' in a strict sense. Instead, it has been made clear that present tense usage in MHG is much more than just a simulation of temporal simultaneity. Present metanarrative devices presuppose the narrator's attitude and thus contribute to establishing a shared experience in the sense of a *common ground*. In this respect, the present tense foregrounds the relationship between the 'teller' and the 'told'

91 A. N. Doane, "Oral texts, intertexts and intratexts: Editing old English". In: Jay Clayton and Eric Rothstein (eds.), *Influence and Intertextuality in Literary History*. Madison 1991, 75–113, 81.

within a shared mental world, resp. the epistemological relation between the 'speaker' and 'reality'. The use of the present within the meta-narrative devices can thus be seen as "a linguistic reflex of the creation of presence and of a reality shared between the poet and the audience in the context of the performance".[92]

In the sections above, it has already been shown how this analysis affects the concepts of 'orality' and 'visualization'. Yet, we still have to ask what consequences can be drawn with regard to the crucial question of the volume which is: How can we link oral poetics and cognitive linguistics in a methodologically and theoretically satisfying manner? For a conclusion in this respect, the last section shifts the perspective and discusses the implications concerning the three components of a Theory of Cognitive Oral Poetics, namely its oral, poetic, and cognitive-linguistic aspects.

Cognitive ORAL Poetics

Since 'orality' is a crucial concept within 'oral poetics', a theory of 'orality' is indispensable when talking about 'cognitive oral poetics'. As seen above, this is not an easy task, in particular when considering investigations of ancient 'oral' *texts*. In this respect, it was shown in Section 4 that a binary distinction between 'written' vs. 'oral' is not able to account for the complex dependencies of textual features and 'orality'. Rather, different dimensions of 'orality' have to be distinguished, as 'oral' can mean 'orally composed', 'orally received', 'oral' as language of 'involvement'[93] resp. 'Nähesprache' ('language of proximity')[94] or 'oral' in the sense of an 'oral style' conditioned by the historically variable relationship between 'literacy' and 'orality' in a particular culture.[95] As a result, "a simple dichotomy between orality and literacy breaks down".[96] If we want to compare 'oral' phenomena in different oral traditions in order to address the question about the universal features of 'oral' poetics, it becomes essential to build upon a finer-grained model, which considers both its different dimensions of 'orality' and factors in the historical-cultural, pragmatic, and epistemological presuppositions.

---

92 Bakker, "Discourse and performance", 16.
93 Chafe, "Integration and involvement".
94 Koch and Oesterreicher, "Sprache der Nähe".
95 Cf. also Zeman, "Historische Mündlichkeit".
96 Bakker, *Pointing*, 45.

Cognitive Oral POETICS

Furthermore, it has become obvious that features of 'orality' have to be distinguished from literary and narrative properties of oral epic poetry. With regard to the 'visual present', it has been shown that its function is not so much a feature of 'orality' (in the strict sense of 'medial orality'), but of performed narrativity. The establishment of a 'shared experience', described through *common ground*$_{1\text{-}3}$ is neither directly linked to the (medially) oral composition of the poem nor to the simultaneity of production and reception, but acts on the performative level of 'oral' epic poems. In this regard, Fleischman has argued that the crucial opposition with regard to deictic means is not the medial distinction between 'spoken' and 'written', but lies within the different pragmatic presuppositions with respect to the (simulated) conversational situation:[97]

> [...] my own probings into this matter suggest that with regard to performed stories the crucial opposition is not the one she invokes between "spoken" and "literary" language—where does "oral literature" fit in?—but between narrative designed as absent-author communication (most varieties of written narrative) and narrative designed for interactive oral performance.[98]

Although poetics and 'orality' are intricately intertwined with respect to oral epic poems, it is therefore necessary that, methodologically, both aspects have to be kept apart when analyzing linguistic phenomena.

COGNITIVE Oral Poetics

Cognitive linguistics, as laid out before, is concerned with the search for universal principles underlying the production and reception of language. Having this aim in mind, it seems to be obvious that cross-linguistic comparisons are inevitable in order to avoid prejudging principles of oral poetics as universal. Based on a typological view on the present tense usage in different oral traditions, it has been pointed out that formal devices alone are not suitable reference points for cross-linguistic comparisons. In consequence, it is thus necessary to look behind the textual surface, in order to uncover the underlying basic functional patterns which can serve as *tertia comparationis*.

97 Fleischman, "Evaluation in narrative", 205.
98 Idem, *Tense and narrativity*, 90.

In respect to the latter, the previous considerations have put emphasis on the concept of *common ground* as one crucial parameter for typological investigations of oral poetics. At the same time, the concept of *common ground* leads us to the very heart of cognitive linguistics: Cognitive semantics "starts from the supposition that meaning resides in *conceptualization*".[99] In this respect, linguistic meanings are not seen as determined solely by objective properties of the situations described, but by "how those situations are apprehended and viewed for linguistic purposes".[100] Perspectivization, i.e. the ability to construe the same situations in alternate ways can thus be seen as a main function of grammar.[101] In selecting and modifying accessible (virtual) worlds, the use of grammatical means thus reflects the relationship between the 'speaker' and the world on the microlinguistic level. Thereby, perspectivization is crucially conditioned by the established origin of perspective and, linked with that, the (different levels of) *common ground* shared by the communicators.

Against this background, grammatical resp. deictic linguistic phenomena ('grounding predications' in terms of Cognitive Grammar) could serve as crucial objects of investigations for looking behind the formal surface on the basic functions of oral poetics characteristics. Not only could this lead us to a better understanding of oral poetics, but it could also offer us a methodological window on what lies within the centre of cognitive oral poetics' interest: The constructing mind that stands behind the poem.

References

Àgel, Vilmos and Mathilde Hennig (eds.). *Grammatik aus Nähe und Distanz. Theorie und Praxis am Beispiel von Nähetexten 1650–2000.* Tübingen 2006.

Amodio, Mark C. *Writing the oral tradition. Oral Poetics and Literate Culture in Medieval England.* Notre Dame, IN 2004.

Auer, Peter. "On line-Syntax—Oder: was es bedeuten könnte, die Zeitlichkeit der mündlichen Sprache ernst zu nehmen". *Sprache und Literatur* 85 (2000): 43–56.

Bakker, Egbert J. "Discourse and Performance: Involvement, Visualization and 'Presence' in Homeric Poetry". *Classical Antiquity* 12.1 (1993): 1–29.

Bakker, Egbert J. "Verbal aspect and mimetic description in Thucydides". In: Egbert J. Bakker (ed.), *Grammar as interpretation. Greek literature in its linguistic contexts,* 7–54. Leiden and New York 1997.

99 Langacker, "English present", 46; emphasis in original.

100 Ibid.

101 Cf. e.g. Ronald W. Langacker, *Foundations of Cognitive Grammar.* Stanford 1991; Elisabeth Leiss, *Sprachphilosophie.* Berlin and New York 2009.

Bakker, Egbert J. "*Mimesis* as Performance: Rereading Auerbach's First Chapter". *Poetics Today* 20.1 (1999): 11–26.
Bakker, Egbert J. *Pointing at the past. From Formula to Performance in Homeric Poetics*. Cambridge, MA and London 2005.
Benveniste, Émile. "Les relations de temps dans le verbe français". *Bulletin de la Société de Linguistique de Paris* 54 (1959) : 69–82.
Boezinger, Bruno. *Das historische Präsens in der älteren deutschen Sprache*. Palo Alto, CA 1912.
Bühler, Karl. *Sprachtheorie. Die Darstellungsfunktion der Sprache*. Stuttgart [1934] 1999.
Carruthers, Mary. *The book of memory: a study of memory in medieval culture*. Cambridge [1990] 2008.
Chafe, Wallace. "Integration and Involvement in Speaking, Writing, and Oral Literature". In: Deborah Tannen (ed.), *Spoken and Written Language: Exploring Orality and Literacy*, 35–53. Norwood 1982.
Cotrozzi, Stefano. *Expect the unexpected. Aspects of pragmatic foregrounding in Old Testament narratives*. New York 2010.
Damsteegt, Theo. "The Present Tense and Internal Focalization of Awareness". *Poetics Today* 26.1 (2005): 39–78.
Däumer, Matthias. *Stimme im Raum und Bühne im Kopf. Über das performative Potenzial der höfischen Artusromane*. Bielefeld 2013.
Diewald, Gabriele. *Deixis und Textsorten im Deutschen*. Tübingen 1991.
Doane, A. N. „Oral texts, intertexts and intratexts: Editing old English". In: Jay Clayton and Eric Rothstein (eds.), *Influence and Intertextuality in Literary History*, 75–113. Madison 1991.
Dobozy, Maria. *Re-Membering the Present. The Medieval German Poet-Minstrel in Cultural Context*. Turnhout 2005.
Elmer, David F. "Presentation Formulas in South Slavic Epic Song". *Oral Tradition* 24.1 (2009): 41–59.
Finnegan, Ruth. *Oral literature in Africa*. Cambridge [1970] 2012.
Fleischman, Suzanne. "Evaluation in Narrative: The Present Tense in Medieval 'Performed Stories'". *Yale French Studies* 70 (1986): 199–251.
Fleischman, Suzanne. *Tense and narrativity. From medieval performance to modern fiction*. London 1990.
Fludernik, Monika. "The Historical Present Tense Yet Again. Tense Switching and Narrative Dynamics in Oral and Quasi-Oral Storytelling". *Text* 11 (1991): 365–398.
Fludernik, Monika. "The historical present tense in English literature: an oral pattern and its literary adaptation". *Language and Literature* 17 (1992): 77–107.
Fludernik, Monika. "The diachronization of narratology". *Narrative* 11.3 (2003): 331–348.
Haferland, Harald. *Mündlichkeit, Gedächtnis und Medialität. Heldendichtung im deutschen Mittelalter*. Göttingen 2004.
Hempel, Heinrich. "Vom 'Präsens Historicum' im Deutschen". In: Heinrich M. Heinrichs (ed.), *Kleine Schriften. Zur Vollendung seines 80. Lebensjahres am 27. August 1965*, 422–429. Heidelberg [1937]1966.
Hennig, Mathilde. *Nähe und Distanzierung. Verschriftlichung und Reorganisation des Nähebereichs im Neuhochdeutschen*. Kassel 2009. URL: http://nbn-resolving.de/urn:nbn:de:hebis:34-2010011231675 (04.05.2016)
Herchenbach, Hugo. *Das Präsens historicum im Mittelhochdeutschen*. Berlin 1911.
Jespersen, Otto. *The philosophy of grammar*. Chicago 1924.

Kawashima, Robert S. "What is Narrative Perspective? A Non-historicist Answer". In: Robert S. Kawashima, Gilles Philippe, and Thelma Sowley (eds.), *Phantom sentences. Essays in linguistics and literature presented to Ann Banfield*, 105–126. Bern and New York 2008.

Koch, Peter and Wulf Oesterreicher. "Sprache der Nähe – Sprache der Distanz. Mündlichkeit und Schriftlichkeit im Spannungsfeld von Sprachtheorie und Sprachgeschichte". *Romanistisches Jahrbuch* 36 (1985): 15–43.

Koch, Peter and Wulf Oesterreicher. *Gesprochene Sprache in der Romania*. Berlin and New York 2011.

Langacker, Ronald W. *Foundations of Cognitive Grammar*. Stanford 1991.

Langacker, Ronald W. "The English present. Temporal coincidence vs. epistemic immediacy". In: Frank Brisard and Adeline Patard (eds.), *Cognitive Approaches to Tense, Aspect, and Epistemic Modality*, 45–86. Amsterdam and Philadelphia 2011.

Leiss, Elisabeth. *Artikel und Aspekt. Die grammatischen Muster von Definitheit*. Berlin and New York 2000.

Leiss, Elisabeth. *Sprachphilosophie*. Berlin and New York 2009.

Lord, Albert Bates, *The Singer of Tale*. Ed. by Stephen Mitchell and Gregory Nagy. Cambridge, MA [1960] 2000.

Parry, Milman. "A comparative study of diction as one of the elements of style in Early Greek Epic poetry". In: Adam Parry (ed.), *The making of Homeric Verse: The Collected Papers of Milman Parry*, 421–436. Oxford 1987.

Parry, Adam. "The language of Achilles". *Transactions and Proceedings of the American Philological Association* 87 (1956): 1–7.

Rubin, David C. *Memory in oral traditions. The cognitive psychology of epic, ballads, and counting-out rhymes*. Oxford 1995.

Sakita, Tomoko. *Reporting Discourse, Tense, and Cognition*. Amsterdam 2002.

Sicking, C. M. J.and Peter Stork. "The grammar of the so-called Historical Present in Ancient Greek". In: Egbert J. Bakker (ed.), *Grammar as interpretation. Greek literature in its linguistic contexts*, 131–168. Leiden and New York 1997.

Sprenger, Ulrike. *Praesens historicum und Praeteritum in der altisländischen Saga*. Basel 1951.

Steadman, J. M. "The Origin of the Historical Present in English". *Studies in Philology* 14 (1917): 1–45.

Tristram, Hildegard L. C. "Tense and Time in Early Irish Narrative". *Innsbrucker Beiträge zur Sprachwissenschaft: Vorträge und kleinere Schriften* 32 (1983): 5–37.

Viti, Carlotta. "The non-literal use of tenses in Latin, with particular reference to the praesens historicum". *Revue de Linguistique Latin* 5 (2010): URL: http://www.zora.uzh.ch/43735/ (accessed April 25, 2015).

Wagner, Silvan. "Narrator and narrative space in Middle High German Epic Poetry". In: Natalia Igl and Sonja Zeman (eds.), *Perspectives on narrativity and narrative perspectivization*. Amsterdam and Philadelphia, 115–138.

Wolfson, Nessa. *CHP. The conversational historical present in American English narrative*. Dordrecht 1982.

Zeman, Sonja. *Tempus und "Mündlichkeit" im Mittelhochdeutschen. Zur Interdependenz grammatischer Perspektivensetzung und "Historischer Mündlichkeit" im mittelhochdeutschen Tempussystem*. Berlin and New York 2010.

Zeman, Sonja. "More than "time": The grammaticalization of the German tense system and the 'frame of reference' as a crucial interface between space and time". In: Kasia Jaszczolt and

Luna Filipović (eds.), *Space and Time in Languages and Cultures. Language diversity*, 157–178. Amsterdam and Philadelphia 2012.

Zeman, Sonja. "Vergangenheit als Gegenwart? Zur Diachronie des Historischen Präsens". In: Petra Maria Vogel (ed.), *Sprachwandel und seine Reflexe im Nhd.* (Jahrbuch der Gesellschaft für Sprachgeschichte 4), 235–255. Berlin and New York 2013.

Zeman, Sonja."Historische Mündlichkeit. Empirische Erörterung einer theoretischen Problemlage". *Zeitschrift für Germanistische Linguistik* 41.3 (2013): 1–36.

Zeman, Sonja. "Nähe, Distanz und (Historische) Pragmatik. Oder: Wie 'nah' ist 'Nähesprache'?". In: Mathilde Hennig and Helmuth Feilke (eds.), "Zur Karriere von 'Nähe und Distanz'". Berlin and New York, 263–302.

Bei Fragen zur Produktsicherheit wenden Sie sich bitte an:
If you have any questions regarding product safety,
please contact:

Walter de Gruyter GmbH
Genthiner Straße 13
10785 Berlin
productsafety@degruyterbrill.com